600077052
D0421153

# Solving ODEs with MATLAB

This book is for people who need to solve ordinary differential equations (ODEs), both initial value problems (IVPs) and boundary value problems (BVPs) as well as delay differential equations (DDEs). These topics are usually taught in separate courses of length one semester each, but *Solving ODEs with MATLAB* provides a sound treatment of all three in about 250 pages. The chapters on each of these topics begin with a discussion of "the facts of life" for the problem, mainly by means of examples. Numerical methods for the problem are then developed – but only the methods most widely used. Although the treatment of each method is brief and technical issues are minimized, the issues important in practice and for understanding the codes are discussed. Often solving a real problem is much more than just learning how to call a code. The last part of each chapter is a tutorial that shows how to solve problems by means of small but realistic examples.

### About the Authors

L. F. Shampine is Clements Professor of Applied Mathematics at Southern Methodist University in Dallas, Texas.

I. Gladwell is Professor of Mathematics at Southern Methodist University in Dallas, Texas.

S. Thompson is Professor of Mathematics at Radford University in Radford, Virginia.

This book distills decades of experience helping people solve ODEs. The authors accumulated this experience in industrial and laboratory settings that include NAG (Numerical Algorithms Group), Babcock and Wilcox Company, Oak Ridge National Laboratory, Sandia National Laboratories, and The MathWorks – as well as in academic settings that include the University of Manchester, Radford University, and Southern Methodist University. The authors have contributed to the subject by publishing hundreds of research papers, writing or editing a half-dozen books, editing leading journals, and writing mathematical software that is in wide use. With associates at The MathWorks, Inc., they wrote all the programs for solving ODEs in MATLAB, programs that are the foundation of this book.

WITHDRAWN

# Solving ODEs with MATLAB

L. F. SHAMPINE
*Southern Methodist University*

I. GLADWELL
*Southern Methodist University*

S. THOMPSON
*Radford University*

CAMBRIDGE
UNIVERSITY PRESS

PUBLISHED BY THE PRESS SYNDICATE OF THE UNIVERSITY OF CAMBRIDGE
The Pitt Building, Trumpington Street, Cambridge, United Kingdom

CAMBRIDGE UNIVERSITY PRESS
The Edinburgh Building, Cambridge CB2 2RU, UK
40 West 20th Street, New York, NY 10011-4211, USA
477 Williamstown Road, Port Melbourne, VIC 3207, Australia
Ruiz de Alarcón 13, 28014 Madrid, Spain
Dock House, The Waterfront, Cape Town 8001, South Africa

http://www.cambridge.org

© L. F. Shampine, I. Gladwell, S. Thompson 2003

This book is in copyright. Subject to statutory exception and
to the provisions of relevant collective licensing agreements,
no reproduction of any part may take place without
the written permission of Cambridge University Press.

First published 2003

Printed in the United States of America

*Typeface* Times 10.5/13 pt.    *System* AMS-T$_{\text{E}}$X    [FH]

*A catalog record for this book is available from the British Library.*

*Library of Congress Cataloging in Publication data*
Shampine, Lawrence F.
Solving ODEs with MATLAB / L.F. Shampine, I. Gladwell, S. Thompson.
p.   cm.
Includes bibliographical references and index.
ISBN 0-521-82404-4 – ISBN 0-521-53094-6 (pb.)
1. Differential equations – Numerical solutions – Data processing. 2. MATLAB.
I. Gladwell, I.   II. Thompson, S., 1948–   III. Title.

QA371.5.D37S43   2003
515′.35 – dc21                                                2003041958

ISBN 0 521 82404 4  hardback
ISBN 0 521 53094 6  paperback

# Contents

# Preface

This book is for people who want to solve ordinary differential equations (ODEs), both initial value problems (IVPs) and boundary value problems (BVPs) as well as delay differential equations (DDEs). *Solving ODEs with MATLAB* is a text for a one-semester course for upper-level undergraduates and beginning graduate students in engineering, science, and mathematics. Prerequisites are a first course in the theory of ODEs and a survey course in numerical analysis. Implicit in these prerequisites is some programming experience, preferably in MATLAB, and some elementary matrix theory. *Solving ODEs with MATLAB* is also a reference for professionals in engineering, science, and mathematics. With it they can quickly obtain an understanding of the issues and see example problems solved in detail. They can use the programs supplied with the book as templates.

It is usual to teach the three topics of this book at an advanced level in separate courses of one semester each. *Solving ODEs with MATLAB* provides a sound treatment of all three topics in about 250 pages. This is possible because of the focus and level of the treatment. The book opens with a chapter called *Getting Started*. Next is a chapter on IVPs. These two chapters must be studied in order, but the remaining two chapters (on BVPs and DDEs) are independent of one another. It is easy to cover one of these chapters in a one-semester course, but the preparation and sophistication of the students will determine whether it is possible to do both. The chapter on DDEs can be covered more quickly than the one on BVPs because only one approach is taken up and it is an extension of methods studied in the chapter on IVPs. Each chapter begins with a discussion of the "facts of life" for the problem, mainly by means of examples. Numerical methods for the problem are then developed – but only the methods most widely used. Although the treatment of each method is brief and technical issues are minimized, the issues important in practice are discussed. Often solving a real problem is much more than just learning how to call a code. The last part of the chapter is a tutorial that shows how to solve problems by means of small but realistic examples.

Although quality software in general scientific computing is discussed, all the examples and exercises are solved in MATLAB. This is most advantageous because MATLAB (2000) has become an extremely important problem-solving environment (PSE) for both teaching and research. The solvers of MATLAB are unusually capable. Moreover, they have a common design and "feel" that make it easy to learn how to use them. MATLAB is such a high-level language that programs are short. This makes it possible to provide complete programs in the text for all the examples. The programs are also provided in electronic form so that they can be used conveniently as templates for similar problems. In particular, the student is asked to modify some of these programs in exercises. Graphics are a part of this PSE, so solutions are typically studied by plotting them. MATLAB has some symbolic algebra capabilities by virtue of a Maple kernel (Maple 1998). *Solving ODEs with MATLAB* exploits these capabilities in the analysis and solution of some of the examples and exercises. There is an Instructor's Manual with solutions for all the exercises. Most of these solutions involve a program, which is available to instructors in electronic form.

The first ODE solver of MATLAB was based on a FORTRAN program written by Larry Shampine and H. A. (Buddy) Watts. For MATLAB 5, Cleve Moler initiated a long and productive relationship between Shampine and The MathWorks. A research and development effort by Shampine and Mark Reichelt (1997) resulted in the MATLAB ODE Suite. The ODE Suite has evolved considerably as a result of further work by Shampine, Reichelt, and Jacek Kierzenka (1999) and the evolution of MATLAB itself. In particular, some of the IVP solvers were given the ability to solve differential algebraic equations (DAEs) of index 1 arising from singular mass matrices. Subsequently, Kierzenka and Shampine (2001) added a program for solving BVPs. Most recently, Skip Thompson, Shampine, and Kierzenka added a program for solving DDEs with constant delays (Shampine & Thompson 2001). We mention this history in part to express our gratitude to Cleve, Mark, and Jacek for the opportunity to work with them on software for this premier PSE and also to make clear that we have a unique understanding of the software that underlies *Solving ODEs with MATLAB*.

Each of us has decades of experience solving ODEs in both academic and nonacademic settings. In this we have contributed to the subject well over 200 papers and half a dozen books, but we have long wanted to write a book that makes our experience in advising people on how to solve ODEs available to a wider audience. *Solving ODEs with MATLAB* is the fulfillment of that wish. We appreciate the help provided by many experts who have commented on portions of the manuscript. Wayne Enright and Jacek Kierzenka have been especially helpful.

# Chapter 1

# Getting Started

## 1.1 Introduction

*Ordinary differential equations* (ODEs) are used throughout engineering, mathematics, and science to describe how physical quantities change, so an introductory course on elementary ODEs and their solutions is a standard part of the curriculum in these fields. Such a course provides insight, but the solution techniques discussed are generally unable to deal with the large, complicated, and nonlinear systems of equations seen in practice. This book is about solving ODEs numerically. Each of the authors has decades of experience in both industry and academia helping people like yourself solve problems. We begin in this chapter with a discussion of what is meant by a numerical solution with standard methods and, in particular, of what you can reasonably expect of standard software. In the chapters that follow, we discuss briefly the most popular methods for important classes of ODE problems. Examples are used throughout to show how to solve realistic problems. MATLAB (2000) is used to solve nearly all these problems because it is a very convenient and widely used *problem-solving environment* (PSE) with quality solvers that are exceptionally easy to use. It is also such a high-level programming language that programs are short, making it practical to list complete programs for all the examples. We also include some discussion of software available in other computing environments. Indeed, each of the authors has written ODE solvers widely used in general scientific computing.

An ODE represents a relationship between a function and its derivatives. One such relation taken up early in calculus courses is the linear ordinary differential equation

$$y'(t) = y(t) \tag{1.1}$$

which is to hold for, say, $0 \le t \le 10$. As we learn in a first course, we need more than just an ODE to specify a solution. Often solutions are specified by means of an initial value. For example, there is a unique solution of the ODE (1.1) for which $y(0) = 1$,

1

namely $y(t) = e^t$. This is an example of an *initial value problem* (IVP) for an ODE. Like this example, the IVPs that arise in practice generally have one and only one solution. Sometimes solutions are specified in a more complicated way. This is important in practice, but it is not often discussed in a first course except possibly for the special case of Sturm–Liouville eigenproblems. Suppose that $y(x)$ satisfies the equation

$$y''(x) + y(x) = 0 \tag{1.2}$$

for $0 \leq x \leq b$. When a solution of this ODE is specified by conditions at both ends of the interval such as

$$y(0) = 0, \quad y(b) = 0$$

we speak of a *boundary value problem* (BVP). A Sturm–Liouville eigenproblem like this BVP always has the trivial solution $y(x) \equiv 0$, but for certain values of $b$ there are non-trivial solutions, too. For instance, when $b = 2\pi$, the BVP has infinitely many solutions of the form $y(x) = \alpha \sin(x)$ for any constant $\alpha$. In contrast to IVPs, which usually have a unique solution, the BVPs that arise in practice may have no solution, a unique solution, or more than one solution. If there is more than one solution, there may be a finite number or an infinite number of them.

Equation (1.1) tells us that the rate of change of the solution at time $t$ is equal to the value of the solution then. In many physical situations, the effects of changes to the solution are delayed until a later time. Models of this behavior lead to *delay differential equations* (DDEs). Often the delays are taken to be constant. For example, if the situation modeled by the ODE (1.1) is such that the effect of a change in the solution is delayed by one time unit, then the DDE is

$$y'(t) = y(t - 1) \tag{1.3}$$

for, say, $0 \leq t \leq 10$. This problem resembles an initial value problem for an ODE; when the delays are constant, both the theory of DDEs and their numerical solution can be based on corresponding results for ODEs. There are, however, important differences. For the ODE (1.1), the initial value $y(0) = 1$ is enough to determine the solution, but that cannot be enough for the DDE (1.3). After all, when $t = 0$ we need $y(-1)$ to define $y'(0)$, but this is a value of the solution prior to the initial time. Thus, an initial value problem for the DDE (1.3) involves not just the value of the solution at the starting time but also its *history*. For this example it is easy enough to argue that, if we specify $y(t)$ for $-1 \leq t \leq 0$, then the initial value problem has a unique solution.

This book is about solving initial value problems for ODEs, boundary value problems for ODEs, and initial value problems for a class of DDEs with constant delays. For brevity we refer throughout to these three kinds of problems as IVPs, BVPs, and DDEs. In the rest of this chapter we discuss fundamental issues that are common to all three. Indeed, some are so fundamental that – even if all you want is a little help solving a specific problem – you need to understand them. The IVPs are taken up in Chapter 2, BVPs in Chapter 3, and DDEs in Chapter 4. The IVP chapter comes first because the ideas and the software of that chapter are used later in the book, so some understanding of this material is needed to appreciate the chapters that follow. The chapters on BVPs and DDEs are mutually independent.

It is assumed that you are acquainted with the elements of programming in MATLAB, so we discuss only matters connected with solving ODEs. If you need to supplement your understanding of the language, the PSE itself has good documentation and there are a number of books available that provide more detail. One that we particularly like is the *MATLAB Guide* (Higham & Higham 2000). Most of the programs supplied with *Solving ODEs with MATLAB* plot solutions on the screen in color. Because it was not practical to provide color figures in the book, we modified the output of these programs to show the solutions in monochrome. Version 6.5 (Release 13) of MATLAB is required for Chapter 4, but version 6.1 suffices for the other chapters. Much of the cited software for general scientific computing is available from general-purpose, scientific computing libraries such as NAG (2002), Visual Numerics (IMSL 2002), and Harwell 2000 (*H2KL*), or from the Netlib Repository (*Netlib*). If the source of the software is not immediately obvious, it can be located through the classification system GAMS, the Guide to Available Mathematical Software (*GAMS*).

Numerical methods and the analytical tools of classical applied mathematics are complementary techniques for investigating and undertaking the solution of mathematical problems. You might be able to solve analytically simple equations involving a very few unknowns, especially with the assistance of a PSE for computer algebra like Maple (1998) or Mathematica (Wolfram 1996). All our examples were computed using the Maple kernel provided with the student version of MATLAB or using the Symbolic Toolbox provided with the professional version.

First we observe that even small changes to the equations can complicate greatly the analytical solutions. For example, Maple is used via MATLAB to solve the ODE

$$y' = y^2$$

at the command line by

```
>> y = dsolve('Dy = y^2')

y = -1/(t-C1)
```

(Sometimes we edit output slightly to give a more compact display.) In this general solution C1 is an arbitrary constant. This family of solutions expressed in terms of a familiar function gives us a lot of insight about how solutions behave. If the ODE is changed "slightly" to

$$y' = y^2 + 1$$

then the general solution is found by dsolve to be

```
y = tan(t+C1)
```

This is more complicated because it expresses the solution in terms of a special function, but it is at least a familiar special function and we understand well how it behaves. However, if the ODE is changed to

$$y' = y^2 + t$$

then the general solution found by dsolve is

```
y = (C1*AiryAi(1,-t)+AiryBi(1,-t))/
    (C1*AiryAi(-t)+AiryBi(-t))
```

which in standard mathematical notation is

$$y(t) = \frac{C_1 Ai'(-t) + Bi'(-t)}{C_1 Ai(-t) + Bi(-t)}$$

Here $Ai(t)$ and $Bi(t)$ are Airy functions. (The Maple kernel denotes these functions by AiryAi and AiryBi, cf. mhelp airy; but MATLAB itself uses different names, cf. help airy.) Again C1 is an arbitrary constant. The Airy functions are not so familiar. This solution is useful for studying the behavior of solutions analytically, but we'd need to plot some solutions to gain a sense of how they behave. Changing the ODE to

$$y' = y^2 + t^2$$

changes the general solution found by dsolve to

```
y = -t*(C1*besselj(-3/4,1/2*t^2)+bessely(-3/4,1/2*t^2))/
      (C1*besselj(1/4,1/2*t^2)+bessely(1/4,1/2*t^2))
```

which in standard mathematical notation is

$$y(t) = -t \frac{C_1 J_{-3/4}\left(\frac{t^2}{2}\right) + Y_{-3/4}\left(\frac{t^2}{2}\right)}{C_1 J_{1/4}\left(\frac{t^2}{2}\right) + Y_{1/4}\left(\frac{t^2}{2}\right)}$$

Again the solution is expressed in terms of special functions, but now they are Bessel functions of fractional order. Again, we'd need to plot some solutions to gain insight. These equations are taken up later in Example 2.3.1.

Something different happens if we change the power of $y$:

```
>> y = dsolve('Dy = y^3 + t^2')
Warning: Explicit solution could not be found.
```

This example shows that even simple-looking equations may not have a solution $y(t)$ that can be expressed in terms of familiar functions by Maple. Such examples are not rare, and usually when Maple fails to find an explicit solution it is because none is known. In fact, for a system of ODEs it is rare that an explicit solution can be found.

For these scalar ODEs it was easy to use a computer algebra package to obtain analytical solutions. Let us now consider some of the differences between solving ODEs analytically and numerically. The analytical solutions of the examples provide valuable insight, but to understand them better we'd need to evaluate and plot some particular solutions. For this we'd need to turn to numerical schemes for evaluating the special functions. But if we must use numerical methods for this, why bother solving them analytically at all? A direct numerical solution might be the best way to proceed for a particular IVP, but Airy and Bessel functions incorporate behavior that can be difficult for numerical methods to reproduce – namely, some have singularities and some oscillate very rapidly. If this is true of the solution that interests us or if we are interested in the solution as $t \rightarrow \infty$, then we may not be able to compute the solution numerically in a straightforward way. In effect, the analytical solution isolates the difficulties and we then rely upon the quality of the software for evaluating the special functions to compute an accurate solution. As the examples show, small changes to the ODE can lead to significant changes in the form of the analytical solution, though this may not imply that the behavior of the solution itself changes much. In contrast, there is really no difference solving IVPs numerically for these equations, including the one for which `dsolve` did not produce a solution. This illustrates the most important virtue of numerical methods: they make it easy to solve a large class of problems. Indeed, our considerable experience is that if an IVP arises in a practical situation, most likely you will not be able to solve it analytically yet you will be able to solve it numerically. On the other hand, the analytical solutions of the examples show how they depend on an arbitrary constant `C1`. Because numerical methods solve one problem at a time, it is not easy to determine how solutions depend on parameters. Such insight can be obtained by combining numerical methods with analytical tools such as variational equations and perturbation methods. Another difference between analytical and numerical solutions is that the standard numerical methods of this book apply only to ODEs defined by smooth functions that are to be solved on a finite interval. It is not unusual for physical problems to involve singular points or an infinite interval. Asymptotic expansions are often combined with numerical methods to deal with these difficulties.

In our view, analytical and numerical methods are complementary approaches to solving ODEs. This book is about numerical methods because they are easy to use and broadly applicable, but some kinds of difficulties can be resolved or understood only by analytical means. As a consequence, the chapters that follow feature many examples of using applied mathematics (e.g., asymptotic expansions and perturbation methods) to assist in the numerical solution of ODEs.

# 1.2 Existence, Uniqueness, and Well-Posedness

From the title of this section you might imagine that this is just another example of mathematicians being fussy. But it is not: it is about whether you will be able to solve a problem at all and, if you can, how well. In this book we'll see a good many examples of physical problems that do not have solutions for certain values of parameters. We'll also see physical problems that have more than one solution. Clearly we'll have trouble computing a solution that does not exist, and if there is more than one solution then we'll have trouble computing the "right" one. Although there are mathematical results that guarantee a problem has a solution and only one, there is no substitute for an understanding of the phenomena being modeled.

Existence and uniqueness are much simpler for IVPs than BVPs, and the class of DDEs we consider can be understood in terms of IVPs, so we concentrate here on IVPs and defer to later chapters a fuller discussion of BVPs and DDEs. The vast majority of IVPs that arise in practice can be written as a system of $d$ explicit first-order ODEs:

$$y_1'(t) = f_1(t, y_1(t), y_2(t), \ldots, y_d(t))$$
$$y_2'(t) = f_2(t, y_1(t), y_2(t), \ldots, y_d(t))$$
$$\vdots$$
$$y_d'(t) = f_n(t, y_1(t), y_2(t), \ldots, y_d(t))$$

For brevity we generally write this system in terms of the (column) vectors

$$y(t) = \begin{pmatrix} y_1(t) \\ y_2(t) \\ \vdots \\ y_d(t) \end{pmatrix}, \quad f(t, y(t)) = \begin{pmatrix} f_1(t, y(t)) \\ f_2(t, y(t)) \\ \vdots \\ f_d(t, y(t)) \end{pmatrix}$$

as

$$y'(t) = f(t, y(t)) \tag{1.4}$$

An IVP is specified by giving values of all the solution components at an initial point,

$$y_1(a) = A_1, \ y_2(a) = A_2, \ \ldots, \ y_d(a) = A_d$$

or, in vector notation,

$$y(a) = A = \begin{pmatrix} A_1 \\ A_2 \\ \vdots \\ A_d \end{pmatrix} \tag{1.5}$$

Using vectors, a system of first-order equations resembles a single equation; in fact, the theory is much the same. However, writing problems as first-order systems is not only convenient for the theory, it is critically important in practice. We'll explain this later and show how to do it.

Roughly speaking, if the function $f(t, y)$ is smooth for all values $(t, y)$ in a region $R$ that contains the initial data $(a, A)$, then the IVP comprising the ODE (1.4) and the initial condition (1.5) has a solution and only one. This settles the existence and uniqueness question for most of the IVPs that arise in practice, but we need to expand on the issue of where the solution exists. The solution extends to the boundary of the region $R$, but that is not the same as saying that it exists throughout a given interval $a \le t \le b$ contained in the region $R$. An example makes the point. The IVP

$$y' = y^2, \quad y(0) = 1$$

has a function $f(t, y) = y^2$ that is smooth everywhere; in other words, it is smooth in the region

$$R = \{-\infty < t < \infty, \ -\infty < y < \infty\}$$

Yet the unique solution

$$y(t) = \frac{1}{1 - t}$$

"blows up" as $t \to 1$ and hence does not exist on a whole interval $0 \le t \le 2$ (say) that is entirely contained in $R$. This does not contradict the existence result because as $t \to 1$, the solution approaches the boundary of the region $R$ in the $y$ variable, a boundary that happens to be at infinity. This kind of behavior is not at all unusual for physical problems. Correspondingly, it is usually reasonable to ask that a numerical scheme approximate a solution well until it becomes too large for the arithmetic of the computer used. Exercises 1.2 and 1.3 take up similar cases.

The form of the ODEs (1.4) and the initial condition (1.5) is standard for IVPs, and in Section 1.3 we look at some examples showing how to write problems in this form. Existence and uniqueness is relatively simple for this standard explicit form, but the properties are more difficult to analyze for equations in the implicit form

$$F(t, y(t), y'(t)) = 0$$

Very simple examples show that both existence and uniqueness are problematic for such equations. For instance, the equation

$$(y'(t))^2 + 1 = 0$$

obviously has no (real) solutions. A more substantial example helps make the point. In scientific and engineering applications there is a great deal of interest in how the solutions $y$ of a system of algebraic equations

$$F(y, \lambda) = 0$$

depend on a (scalar) parameter $\lambda$. Differentiating with respect to the parameter, we find that

$$\frac{\partial F}{\partial y}\frac{dy}{d\lambda} + \frac{\partial F}{\partial \lambda} = 0$$

This is a system of first-order ODEs. If for some $\lambda_0$ we can solve the algebraic equations $F(y, \lambda_0) = 0$ for $y(\lambda_0) = y_0$, then this provides an initial condition for an IVP for $y(\lambda)$. If the Jacobian matrix

$$J = \frac{\partial F}{\partial y} = \left(\frac{\partial F_i}{\partial y_j}\right)$$

is nonsingular, we can write the ODEs in the standard form

$$\frac{dy}{d\lambda} = -J^{-1}\frac{\partial F}{\partial \lambda}$$

However, if the Jacobian matrix is singular then the questions of existence and uniqueness are much more difficult to answer. This is a rather special situation, but in fact it is often the situation with the most interesting science. It is when solutions bifurcate – that is, the number of solutions changes. If we are to apply standard codes for IVPs at such a singular (bifurcation) point, we must resort to the analytical tools of applied mathematics to sort out the behavior of solutions near this point. Exercise 1.1 considers a similar problem.

As a concrete example of bifurcation, suppose that we are interested in steady-state (constant) solutions of the ODE

$$y' = y^2 - \lambda$$

The steady states are solutions of the algebraic equation

$$0 = y^2 - \lambda \equiv F(y, \lambda)$$

It is obvious that, for $\lambda \geq 0$, one steady-state solution is $y(\lambda) = \sqrt{\lambda}$. However, to study more generally how the steady state depends on $\lambda$, we could compute it as the solution of the IVP

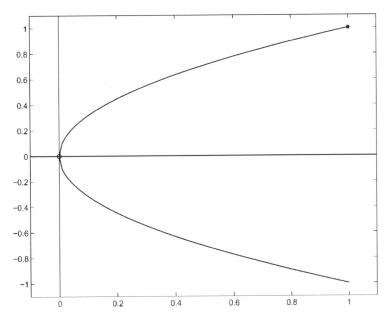

Figure 1.1: $(0, 0)$ is a singular point for $2yy' - 1 = 0$.

$$2y\frac{dy}{d\lambda} - 1 = 0, \quad y(1) = 1$$

Provided that $y \neq 0$, the ODE can be written immediately in standard form and solved for values of $\lambda$ decreasing from 1. However, the equation is singular when $y(\lambda) = 0$, which is true for $\lambda = 0$. The singular point $(0, 0)$ leaves open the possibility that there is more than one solution of the ODE passing through this point, and so there is: $y(\lambda) = -\sqrt{\lambda}$ is a second solution. Using standard software, we can start at $\lambda = 1$ and integrate the equation easily until close to the origin, where we run into trouble because $y'(\lambda) \to \infty$ as $\lambda \to 0$. See Figure 1.1.

For later use in discussing numerical methods, we need to be a little more precise about what we mean by a *smooth* function $f(t, y)$. We mean that it is continuous in a region $R$ and that it has continuous derivatives with respect to the dependent variables there – as many derivatives as necessary for whatever argument we make. A technical condition is that $f$ must satisfy a *Lipschitz condition* in the region $R$. That is, there is a constant $L$ such that, for any points $(t, u)$ and $(t, v)$ in the region $R$,

$$\|f(t, u) - f(t, v)\| \leq L\|u - v\|$$

In the case of a single equation, the mean value theorem states that

$$f(t, u) - f(t, v) = \frac{\partial f}{\partial y}(t, \zeta)(u - v)$$

so $f(t, y)$ satisfies a Lipschitz condition if $\left| \frac{\partial f(t,y)}{\partial y} \right|$ is bounded in the region $R$ by a constant $L$. Similarly, if the first partial derivatives $\left| \frac{\partial f_i(t, y_1, y_2, \ldots, y_d)}{\partial y_j} \right|$ are all bounded in the region $R$, then the vector function $f(t, y)$ satisfies a Lipschitz condition there.

Roughly speaking, a *well-posed problem* is one for which small changes to the data lead to small changes in the solution. Such a problem is also said to be *well-conditioned* with respect to changes in the data. This is a fundamental property of a physical problem and it is also fundamental to the numerical solution of the problem. The methods that we study can be regarded as producing the exact solution to a problem with the data that defines the problem changed a little. For a well-posed problem, this means that the numerical solution is close to the solution of the given problem. In practice this is all blurred because it depends both on how much accuracy you want in a solution and on the arithmetic you use in computing it. Let's now discuss a familiar example that illuminates some of the issues.

Imagine that we have a pendulum: a light, rigid rod hanging vertically from a frictionless pivot with a heavy weight (the bob) at the free end. With a particular choice of units, the angle $\theta(t)$ that the pendulum makes with the vertical at time $t$ satisfies the ODE

$$\theta'' + \sin(\theta) = 0 \tag{1.6}$$

Suppose that the pendulum is hanging vertically so that the initial angle $\theta(0) = 0$ and that we thump the bob to give it an initial velocity $\theta'(0)$. When the initial velocity is zero, the pendulum does not move at all. If the velocity is nonzero and small enough, the pendulum will swing back and forth. Figure 1.2 shows $\theta(t)$ for several such solutions, namely those with initial velocities $\theta'(0) = -1.9$, 1.5, and 1.9. There is another kind of solution. If we thump the bob hard enough, the pendulum will swing over the top and, with no friction, it will whirl around the pivot forever. This is to say that if the initial velocity $\theta'(0)$ is large enough then $\theta(t)$ will increase forever. The figure shows two such solutions with initial velocities $\theta'(0) = 2.1$ and 2.5. If you think about it, you'll realize that there is a very special solution that occurs as the solutions change from oscillatory to increasing. This solution is the dotted curve in Figure 1.2. Physically, it corresponds to an initial velocity that causes the pendulum to approach and then come to rest vertically and upside down. Clearly this solution is unstable – an arbitrarily small change to the initial velocity gives rise to a solution that is eventually very different. In other words, the IVP for this initial velocity is ill-posed (ill-conditioned) on long time intervals.

Interestingly, we can deduce the initial velocity that results in the unstable solution of (1.6). This is a conservative system, meaning that the *energy*

$$E(t) = 0.5(\theta'(t))^2 - \cos(\theta(t))$$

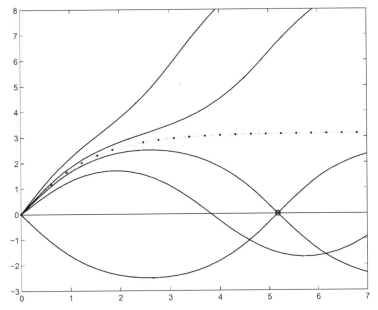

Figure 1.2: $\theta(t)$, the angle from the vertical of the pendulum.

is constant. To prove this, differentiate the expression for $E(t)$ and use the fact that $\theta(t)$ satisfies the ODE (1.6) to see that the derivative $E'(t)$ is zero for all $t$. On physical grounds, the solution of interest satisfies the condition $\theta(\infty) = \pi$ and, a fortiori, $\theta'(\infty) = 0$. Along with the initial value $\theta(0) = 0$, conservation of energy tells us that for this solution

$$0.5 \times (\theta'(0))^2 - \cos(0) = 0.5 \times 0^2 - \cos(\pi)$$

and hence that $\theta'(0) = 2$. With this we have the unstable solution defined as the solution of the IVP consisting of equation (1.6) and initial values $\theta(0) = 0$ and $\theta'(0) = 2$. The other solutions of Figure 1.2 were computed using the MATLAB IVP solver $\texttt{ode45}$ and default error tolerances, but these tolerances are not sufficiently stringent to compute an accurate solution of the unstable IVP.

The unstable solution is naturally described as the solution of a boundary value problem. It is the solution of the ODE (1.6) with boundary conditions

$$\theta(0) = 0, \quad \theta(\infty) = \pi \tag{1.7}$$

When modeling a physical situation with a BVP, it is not always clear what boundary conditions to use. We have already commented that, on physical grounds, $\theta'(\infty) = 0$ also. Should we add this boundary condition to (1.7)? No; just as with IVPs, two conditions are needed to specify the solution of a second-order equation and three are too many. But should we use this boundary condition at infinity or should we use $\theta(\infty) = \pi$? A clear difficulty is that, in addition to the solution $\theta(t)$ that we want, the BVP with boundary condition $\theta'(\infty) = 0$ has (at least) two other solutions, namely $-\theta(t)$ and $\theta(t) \equiv 0$. We computed the unstable solution of Figure 1.2 by solving the BVP (1.6) and (1.7) with the MATLAB BVP solver bvp4c. The solution of the BVP is well-posed, so we could use the default error tolerances. On the other hand, the BVP is posed on an infinite interval, which presents its own difficulties. All the codes we discuss in this book are intended for problems defined on finite intervals. As we see here, it is not unusual for physical problems to be defined on infinite intervals. Existence, uniqueness, and well-posedness are not so clear then. One approach to solving such a problem, which we actually used for the figure, follows the usual physical argument of imposing the conditions at a finite point so distant that it is idealized as being at infinity. For the figure, we solved the ODE subject to the boundary conditions

$$\theta(0) = 0, \quad \theta(100) = \pi$$

It turned out that taking the interval as large as [0, 100] was unnecessarily cautious because the steady state of $\theta$ is almost achieved for $t$ as small as 7. For the BVP (1.6) and (1.7), we can use the result $\theta'(0) = 2$ derived earlier as a check on the numerical solution and in particular to check whether the interval is long enough. With default error tolerances, bvp4c produces a numerical solution that has an initial slope of $\theta'(0) = 1.999979$, which is certainly good enough for plotting the figure.

Another physical example shows that some BVPs do not have solutions and others have more than one. The equations

$$y' = \tan(\phi)$$

$$v' = -\frac{g \sin(\phi) + \nu v^2}{v \cos(\phi)} \tag{1.8}$$

$$\phi' = -\frac{g}{v^2}$$

describe a projectile problem, the planar motion of a shot fired from a cannon. Here the solution component $y$ is the height of the shot above the level of the cannon, $v$ is the velocity of the shot, and $\phi$ is the angle (in radians) of the trajectory of the shot with the horizontal. The independent variable $x$ measures the horizontal distance from the cannon. The constant $\nu$ represents air resistance (friction) and $g = 0.032$ is the appropriately scaled gravitational constant. These equations neglect three-dimensional effects such as cross winds and rotation of the shot. The initial height is $y(0) = 0$ and there is a given muzzle velocity $v(0)$ for the cannon. The standard projectile problem is to choose the initial angle $\phi(0)$ of the cannon (and hence of the shot) so that the shot will hit a target at the same height as the cannon at distance $x = x_{\text{end}}$. That is, we require $y(x_{\text{end}}) = 0$. All together, the boundary conditions are

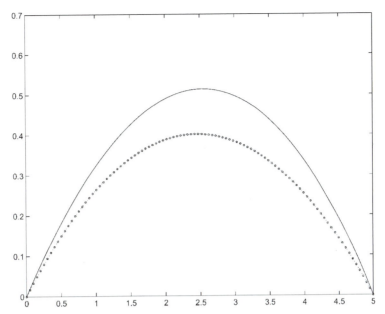

Figure 1.3: Two ways to hit a target at $x_{end} = 5$ when $v(0) = 0.5$ and $v = 0.02$.

$$y(0) = y(x_{end}) = 0, \quad v(0) \text{ given}$$

Notice that we specify three boundary conditions. Just as with IVPs, for a system of three first-order equations we need three boundary conditions to determine a solution. Does this boundary value problem have a solution? It certainly does not for $x_{end}$ beyond the range of the cannon. On the other hand, if $x_{end}$ is small enough then we expect a solution, but is there only one? No, suppose that the target is close to the cannon. We can hit it by shooting with an almost flat trajectory or by shooting high and dropping the shot on the target. That is, there are (at least) two solutions that correspond to initial angles $\phi(0) = \phi_{low} \approx 0$ and $\phi(0) = \phi_{high} \approx \pi/2$. As it turns out, there are exactly two solutions. Now, let $x_{end}$ increase. There are still two solutions, but the larger the value of $x_{end}$, the smaller the angle $\phi_{high}$ and the larger the angle $\phi_{low}$. Figure 1.3 shows such a pair of trajectories. If we keep increasing $x_{end}$, eventually we reach the maximum distance possible with the given muzzle velocity. At this distance there is just one solution, $\phi_{low} = \phi_{high}$. In summary, there is a critical value of $x_{end}$ for which there is exactly one solution. If $x_{end}$ is smaller than this critical value then there are exactly two solutions; if it is larger, there is no solution at all.

For IVPs we have an existence and uniqueness result that deals with most of the problems that arise physically. There are mathematical results that assert existence and say something about the number of solutions of BVPs, but they are so special that they are seldom important in practice. Instead you must rely on your understanding of the problem to have any confidence that it has a solution and is well-posed. Determining the number of solutions is even more difficult, and in practice about the best we can do is look for a solution close to a guess. There is a real possibility of computing a "wrong" solution or a solution with unexpected behavior.

Stability is the key to understanding numerical methods for the solution of IVPs defined by equation (1.4) and initial values (1.5). All the methods that we study produce approximations $y_n \approx y(t_n)$ on a mesh

$$a = t_0 < t_1 < t_2 < \cdots < t_N = b \tag{1.9}$$

that is chosen by the algorithm. The integration starts with the given initial value $y_0 = y(a) = A$ and, on reaching $t_n$ with $y_n \approx y(t_n)$, the solver computes an approximation at $t_{n+1} = t_n + h_n$. The quantity $h_n$ is called the *step size,* and computing $y_{n+1}$ is described as taking a step from $t_n$ to $t_{n+1}$.

What the solver does in taking a step is not what you might expect. The *local solution* $u(t)$ is the solution of the IVP

$$u' = f(t, u), \quad u(t_n) = y_n \tag{1.10}$$

In taking a step, the solver tries to find $y_{n+1}$ so that the *local error*

$$u(t_{n+1}) - y_{n+1}$$

is no larger than error tolerances specified by the user. This controls the *true (global) error*

$$y(t_{n+1}) - y_{n+1}$$

only indirectly. The propagation of error can be understood by writing the error at $t_{n+1}$ as

$$y(t_{n+1}) - y_{n+1} = [u(t_{n+1}) - y_{n+1}] + [y(t_{n+1}) - u(t_{n+1})]$$

The first term on the right is the local error, which is controlled by the solver. The second is the difference at $t_{n+1}$ of two solutions of the ODE that differ by $y(t_n) - y_n$ at $t_n$. It is a characteristic of the ODE and hence cannot be controlled directly by the numerical method. If the IVP is unstable – meaning that some solutions of the ODEs starting near $y(t)$ spread apart rapidly – then we see from this that the true errors can grow even when the local errors are small at each step. On the other hand, if the IVP is stable so that solutions come together, then the true errors will be comparable to the local errors. Figure 1.2 shows what can happen. As a solver tries to follow the unstable solution plotted with dots, it makes small errors that move the numerical solution on to nearby solution curves. As the figure makes clear, local solutions that start near the unstable solution spread out; the cumulative effect is a very inaccurate numerical solution, even when the solver is able to follow closely each local solution over the span of a single step. It is very important to understand this view of numerical error, for it makes clear a fundamental limitation on all the numerical methods that we consider. No matter how good a job the numerical method does in approximating the solution over the span of a step, if the IVP is unstable then you will eventually compute numerical solutions $y_j$ that are far from the desired solution values $y(t_j)$. How quickly this happens depends on how accurately the method tracks the local solutions and how unstable the IVP is.

A simple example will help us understand the role of stability. The solution of the ODE

$$y' = 5(y - t^2) \tag{1.11}$$

with initial value $y(0) = 0.08$ is

$$y(t) = t^2 + 0.4t + 0.08$$

The IVP and its solution seem innocuous, but the general solution of the ODE is

$$(t^2 + 0.4t + 0.08) + Ce^{5t} \tag{1.12}$$

for an arbitrary constant $C$. The ODE is unstable because a solution with $C = C_1$ and a solution with $C = C_2$ differ by $(C_1 - C_2)e^{5t}$, a quantity that grows exponentially fast in time. To understand what this means for numerical solution of the IVP, suppose that in the first step we make a small local error so that $y_1$ is not exactly equal to $y(t_1)$. In the next step we try to approximate the local solution $u(t)$ defined by the ODE and the initial condition $u(t_1) = y_1$. It has the form (1.12) with a small nonzero value of $C$ determined by the initial condition. Suppose that we make no further local errors, so that we compute $y_n = u(t_n)$ for $n = 2, 3, \ldots$. The true error then is $y(t_n) - u(t_n) = Ce^{5t_n}$. No matter how small the error in the first step, before long the exponential growth of the true error will result in an unacceptable numerical solution $y_n$.

For the example of Figure 1.2, the solution curves come together when we integrate from right to left, which is to say that the dotted solution curve is stable in that direction. Sometimes we have a choice of direction of integration, and it is important to appreciate that the stability of IVPs may depend on this direction. The direction field and solution curves for the ODE

$$y' = \cos(t)y \tag{1.13}$$

displayed in Figure 1.4 are illuminating. In portions of the interval, solutions of the ODE spread apart; hence the equation is modestly unstable there. In other portions of the interval, solutions of the ODE come together and the equation is modestly stable. For this equation, the direction of integration is immaterial. This example shows that it is an oversimplification to say simply that an IVP is unstable or stable. Likewise, the growth or decay of errors made at each step by a solver can be complex. In particular, you should

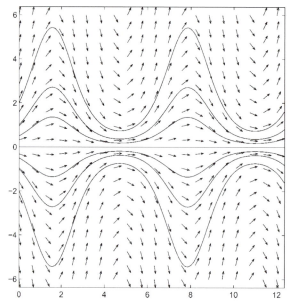

Figure 1.4:  Direction field and solutions of the ODE
$y' = \cos(t)y$.

not assume that errors always accumulate. For systems of ODEs, one component of the solution can be stable and another unstable at the same time. The coupling of the components of a system can make the overall behavior unclear.

A numerical experiment shows what can happen. Euler's method is a basic scheme discussed fully in the next chapter. It advances the numerical solution of $y' = f(t, y)$ a distance $h$ using the formula

$$y_{n+1} = y_n + hf(t_n, y_n) \tag{1.14}$$

The solution of the ODE (1.13) with initial value $y(0) = 2$ is

$$y(t) = 2e^{\sin(t)}$$

The local solution $u(t)$ is the solution of (1.13) that goes through the point $(t_n, y_n)$, namely

$$u(t) = y_n e^{(\sin(t) - \sin(t_n))}$$

Figure 1.5 shows the local and global errors when a constant step size of $h = 0.1$ is used to integrate from $t = 0$ to $t = 3$. Although we are not trying to control the size of the local errors, they do not vary greatly. By definition, the local and global errors are the same in

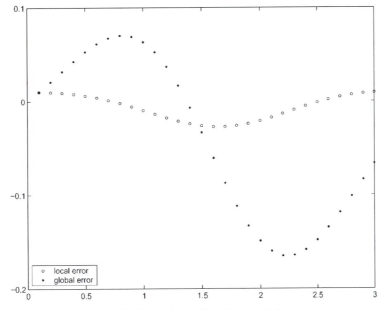

Figure 1.5: Comparison of local and global errors.

the first step. Thereafter, the global errors grow and decay according the stability of the problem, as seen in Figure 1.4.

*Backward error analysis* has been an invaluable tool for understanding issues arising in numerical linear algebra. It provides a complementary view of numerical methods for ODEs that is especially important for the methods of the MATLAB solvers. All these solvers produce approximate solutions $S(t)$ on the whole interval $[a, b]$ that are piecewise smooth. For conceptual purposes, we can define a piecewise-smooth function $S(t)$ with $S(t_n) = y_n$ for each value $n$ that plays the same role for methods that do not naturally produce such an approximation. The *residual* of such an approximation is

$$r(t) = S'(t) - f(t, S(t))$$

Put differently, $S(t)$ is the *exact* solution of the perturbed ODE

$$S'(t) = f(t, S(t)) + r(t)$$

In the view of backward error analysis, $S(t)$ is a good approximate solution if it satisfies an ODE that is "close" to the one given – that is, if the residual $r(t)$ is "small". This is a perfectly reasonable definition of a "good" solution, but if the IVP is well-posed then it also implies that $S(t)$ is "close" to the true solution $y(t)$, the usual definition of a good approximation. In this view, a solver tries to produce an approximate solution with a small residual. The BVP solver of MATLAB does exactly this, and the IVP and DDE solvers do it indirectly.

### ■ EXERCISE 1.1

Among the examples available through the MATLAB command `help dsolve` is the IVP

$$(y')^2 + y^2 = 1, \quad y(0) = 0$$

In addition to showing how easy it is to solve simple IVPs analytically, the example has interesting output:

```
>> y = dsolve('(Dy)^2 + y^2 = 1','y(0) = 0')

y = [  sin(t)]
    [ -sin(t)]
```

According to `dsolve`, this IVP has two solutions. Is this correct? If it is, reconcile this with the existence and uniqueness result for IVPs of Section 1.2.

### ■ EXERCISE 1.2

Prove that the function $f(t, y)$ in

$$y' = f(t, y) = +\sqrt{|y|}$$

does not satisfy a Lipschitz condition on the rectangle $|t| \leq 1$, $|y| \leq 1$. Show by example that this ODE has more than one solution that satisfies $y(0) = 0$. Show that $f(t, y)$ does satisfy a Lipschitz condition on the rectangle $|t| \leq 1$, $0 < \alpha \leq y \leq 1$. The general result discussed in the text then says that the ODE has only one solution with its initial value in this rectangle.

### ■ EXERCISE 1.3

The interval on which the solution of an IVP exists depends on the initial conditions. To see this, find the general solution of the following ODEs and consider how the interval of existence depends on the initial condition:

$$y' = \frac{1}{(t-1)(t-2)}$$
$$y' = -3y^{4/3} \sin(t)$$

### ■ EXERCISE 1.4

The program `dfs.m` that accompanies this book provides a modest capability for computing a direction field and solutions of a scalar ODE, $y' = f(t, y)$. The first argument of `dfs.m` is a string defining $f(t, y)$. In this the independent variable must be called $t$ and the dependent variable must be called $y$. The second argument is an array `[wL wR wB wT]` specifying a plot window. Specifically, solutions are plotted for values $y(t)$ with $wL \leq t \leq wR$, $wB \leq y \leq wT$. The program first plots a direction field. If you then indicate a point in the plot window by placing the cursor there and clicking, it computes and plots the solution of the ODE through this point. Clicking at a point outside the window terminates the run. For example, Figure 1.4 can be reproduced with the command

```
>> dfs('cos(t)*y',[0 12 -6 6]);
```

and clicking at appropriate points in the window. Use `dfs.m` to study graphically the stability of the ODE (1.11). A plot window appropriate for the IVP studied analytically in the text is given by `[0 5 -2 20]`.

### ■ EXERCISE 1.5

Compare local and global errors as in Figure 1.5 when solving equation (1.11) with $y(0) = 0.08$. Use Euler's method with the constant step size $h = 0.1$ to integrate from 0 to 2. The stability of this problem is studied analytically in the text and numerically in Exercise 1.4. With this in mind, discuss the behavior of the global errors.

# 1.3 Standard Form

Ordinary differential equations arise in the most diverse forms. In order to solve an ODE problem, you must first write it in a form acceptable to your code. By far the most common form accepted by IVP solvers is the system of first-order equations discussed in Section 1.2,

$$y' = f(t, y) \tag{1.15}$$

The MATLAB IVP solvers accept ODEs of the more general form

$$M(t, y)y' = F(t, y) \tag{1.16}$$

involving a nonsingular *mass matrix* $M(t, y)$. These equations can be written in the form (1.15) with $f(t, y) = M(t, y)^{-1}F(t, y)$, but for some kinds of problems the form (1.16) is more convenient and more efficient. With either form, we must formulate the ODEs as a system of first-order equations. The usual way to do this is to introduce new dependent variables. You must introduce a new variable for each of the dependent variables in the original form of the problem. In addition, a new variable is needed for each derivative of an original variable up to one less than the highest derivative appearing in the original equations. For each new variable, you need an equation for its first derivative expressed in terms of the new variables. A little manipulation using the definitions of the new variables and the original equations is then required to write the new equations in the form (1.15) (or (1.16)). This is harder to explain in words than it is to do, so let's look at some examples. To put the ODE (1.6) describing the motion of a pendulum in standard form, we begin with a new variable $y_1(t) = \theta(t)$. The second derivative of $\theta(t)$ appears in the equation, so we need to introduce one more new variable, $y_2(t) = \theta'(t)$. For these variables we have

$$y_1'(t) = \theta'(t) = y_2(t)$$
$$y_2'(t) = \theta''(t) = -\sin(\theta(t)) = -\sin(y_1(t))$$

From this we recognize that

$$y_1' = y_2$$
$$y_2' = -\sin(y_1)$$

that is, the two components of the vector function $f(t, y)$ of (1.15) are given by $f_1(t, y) = y_2$ and $f_2(t, y) = -\sin(y_1)$. When we solved an IVP for this ODE we specified initial values

$$y_1(0) = \theta(0) = 0$$
$$y_2(0) = \theta'(0)$$

and when we solved a BVP we specified boundary values

$$y_1(0) = \theta(0) = 0$$
$$y_1(b) = \theta(b) = \pi$$

As another example consider Kepler's equations describing the motion of one body around another of equal mass located at the origin under the influence of gravity. In appropriate units they have the form

$$x'' = -\frac{x}{r^3}, \quad y'' = -\frac{y}{r^3} \tag{1.17}$$

where $r = \sqrt{x^2 + y^2}$. Here $(x(t), y(t))$ are the coordinates of the moving body relative to the body fixed at the origin. With initial values

$$x(0) = 1 - e, \quad y(0) = 0, \quad x'(0) = 0, \quad y'(0) = \sqrt{\frac{1+e}{1-e}} \tag{1.18}$$

there is an analytical solution in terms of solutions of Kepler's (algebraic) equation that shows the orbit is an ellipse of eccentricity $e$. These equations are easily written as a first-order system. One choice is to introduce variables $y_1 = x$ and $y_2 = y$ for the unknowns and then, because the second derivatives of the unknowns appear in the equations, to introduce variables $y_3 = x'$ and $y_4 = y'$ for their first derivatives. You should verify that the first-order system is

$$y_1' = y_3$$
$$y_2' = y_4$$
$$y_3' = -\frac{y_1}{r^3}$$
$$y_4' = -\frac{y_2}{r^3}$$

where $r = \sqrt{y_1^2 + y_2^2}$, and that the initial conditions are

$$y_1(0) = 1 - e, \quad y_2(0) = 0, \quad y_3(0) = 0, \quad y_4(0) = \sqrt{\frac{1+e}{1-e}}$$

Both of these examples illustrate the fact that mechanical problems described by Newton's laws of motion lead to systems of second-order equations and, if there is no dissipation, there are no first derivatives. Equations like this are called *special second-order equations*. They are sufficiently common that some codes accept IVPs in the standard form

$$y'' = f(t, y)$$

with initial position $y(a)$ and initial velocity $y'(a)$ given. As we have seen, it is easy enough to write such problems as first-order systems, but since there are numerical methods that take advantage of the special form it is both efficient and convenient to work directly with the system of second-order equations (cf. Brankin et al. 1989).

Sometimes it is useful to introduce additional unknowns in order to compute quantities related to the solution. An example arises in formulating the solution of the Sturm–Liouville eigenproblem consisting of the ODE

$$y''(x) + \lambda y(x) = 0$$

with boundary conditions $y(0) = 0$ and $y(2\pi) = 0$. The task is to find an eigenvalue $\lambda$ for which there is a nontrivial (i.e., not identically zero) solution, known as an eigenfunction. For some purposes it is appropriate to normalize the solution so that

$$1 = \int_0^{2\pi} y^2(t)\,dt$$

A convenient way to impose this normalizing condition is to introduce a variable

$$y_3(x) = \int_0^x y^2(t) \, dt$$

Then, along with variables $y_1(x) = y(x)$ and $y_2(x) = y'(x)$, we have the first-order system

$$y_1' = y_2$$
$$y_2' = -\lambda y_1$$
$$y_3' = y_1^2$$

The definition of the new variable implies that $y_3(0) = 0$, and we seek a solution of the system of ODEs for which $y_3(2\pi) = 1$. All together we have three equations and one unknown parameter $\lambda$. The solution of interest is to be determined by the four boundary conditions

$$y_1(0) = 0, \quad y_1(2\pi) = 0, \quad y_3(0) = 0, \quad y_3(2\pi) = 1$$

Here we use the device of introducing a new variable for an auxiliary quantity to determine a solution of interest. Another application is to put the problem in standard form. The MATLAB BVP solver `bvp4c` accepts problems with unknown parameters, but this facility is not commonly available. Most BVP solvers require that the parameter $\lambda$ be replaced by a variable $y_4(t)$. The parameter is constant, so the new unknown satisfies the ODE

$$y_4' = 0$$

In this way we obtain a system of four first-order ODEs that does not explicitly involve an unknown parameter,

$$y_1' = y_2$$
$$y_2' = -y_4 y_1$$
$$y_3' = y_1^2$$
$$y_4' = 0$$

and the boundary conditions are unchanged. Exercises 1.8 and 1.9 exploit this technique of converting integral constraints to differential equations.

Often a proper selection of unknowns is key to solving a problem. The following example arose in an investigation by chemical engineer F. Song (pers. commun.) into the corrosion of natural gas pipelines under a coating with cathodic protection. The equations are naturally formulated as

$$\frac{d^2x}{dz^2} = \gamma(e^x + \mu_c e^{x\omega_{Fe}} + \lambda_H e^{x\omega_H} + \lambda_{O_2} e^{x\omega_{O_2}})$$

$$\frac{d^2p_{O_2}}{dz^2} = \pi p_{O_2} e^{x\omega_{O_2}} + \beta p_{O_2} + \kappa$$

This is a BVP with boundary conditions at the origin and infinity. It is possible to eliminate the variable $p_{O_2}(z)$ to obtain a fourth-order equation for the solution variable $x(z)$ alone. Reducing a set of ODEs to a single, higher-order equation is often useful for analysis, but to solve the problem numerically the equation must then be reformulated as a system of first-order equations. If you forget about the origin of the fourth-order ODE for $x(z)$ here, you might reasonably introduce new variables in the usual way,

$$y_1 = x, \quad y_2 = x', \quad y_3 = x'', \quad y_4 = x'''$$

This is not a good idea because it does not directly account for the behavior of the corrodant, $p_{O_2}(z)$. It is much better practice here to start with the original formulation and introduce the new variables

$$w_1 = x, \quad w_2 = x', \quad w_3 = p_{O_2}, \quad w_4 = p'_{O_2}$$

It is easier to select appropriate error tolerances for quantities that can be interpreted physically. Also, by specifying error tolerances for $w_3$, we require the solver to compute accurately the fundamental quantity $p_{O_2}$. When solving BVPs you must provide a guess for the solution. It is easier to provide a reasonable guess for quantities that have physical significance. In Song's work, a suitable formulation of this problem and a corresponding guess was important to the successful solution of this BVP. It is worth noting that here "solving" the problem was not just a matter of computing the solution of a single BVP. As is so often the case in practice, the BVP was to be solved for a range of parameter values.

### ■ EXERCISE 1.6

Consider the two-point BVP consisting of the second-order ODE

$$(p(x)y')' + q(x)y = r(x)$$

with boundary conditions

$$y(0) = 0, \quad p(1)y'(1) = 2$$

The function $p(x)$ is differentiable and positive for all $x \in [0, 1]$. Using $p'(x)$, write this problem in the form of a first-order system using as unknowns $y_1 = y$ and $y_2 = y'$. In applications it is often natural to use the flux $py'$ as an unknown instead of $y'$. Indeed, one of the boundary conditions here states that the flux has a given value. Show that with the flux as an unknown, you can write the problem in the form of a first-order system without needing to differentiate $p(x)$.

■  **EXERCISE 1.7**

Kamke (1971, p. 598) states that the IVP

$$y(y'')^2 = e^{2x}, \quad y(0) = 0, \ y'(0) = 0$$

describes space charge current in a cylindrical capacitor.

- Find two equivalent explicit ODEs in special second-order form.
- Formulate the second-order equations as systems of first-order equations.

■  **EXERCISE 1.8**

Murphy (1965) extends the classical Falkner–Skan similarity solutions for laminar incompressible boundary layer flows to flows over curved surfaces. He derives a BVP consisting of the ODE

$$f'''' + (\Omega + f)f''' + \Omega f f'' - (2\beta - 1)[f'f'' + \Omega(f')^2] = 0$$

to be solved on $0 \le \eta \le b$ with boundary conditions

$$f(0) = f'(0) = 0, \quad f'(b) = e^{-\Omega b}, \quad f''(b) = -\Omega e^{-\Omega b}$$

Here $\Omega$ is a curvature parameter, $\beta$ is a pressure–gradient parameter, and $b$ is large enough that the exponential terms in the boundary conditions describe the correct asymptotic behavior. Physically significant quantities are the displacement thickness

$$\Delta^* = \int_0^b [1 - f'(\eta)e^{\Omega\eta}] \, d\eta$$

and the momentum thickness

$$\theta = \int_0^b f'(\eta)e^{\Omega\eta}[1 - f'(\eta)e^{\Omega\eta}] \, d\eta$$

Formulate the BVP in terms of a system of first-order equations. Add equations and initial values so that the displacement thickness and the momentum thickness can each be computed along with the solution $f(\eta)$.

■  **EXERCISE 1.9**

Caughy (1970) describes the large-amplitude whirling of an elastic string by a BVP consisting of the ODE

$$\mu'' + \omega^2 \left( \frac{1 - \alpha^2}{H} \frac{1}{\sqrt{1 + \mu^2}} + \alpha^2 \right) \mu = 0$$

and boundary conditions

$$\mu'(0) = 0, \quad \mu'(1) = 0$$

Here $\alpha$ is a physical constant with $0 < \alpha < 1$. Because the whirling frequency $\omega$ is to be determined as part of solving the BVP, there must be another boundary condition. Caughy specifies the amplitude $\varepsilon$ of the solution at the origin:

$$\mu(0) = \varepsilon$$

An unusual aspect of this problem is that an important constant $H$ is defined in terms of the solution $\mu(x)$ throughout the interval of integration:

$$H = \frac{1}{\alpha^2} \left[ 1 - (1 - \alpha^2) \int_0^1 \frac{dx}{\sqrt{1 + \mu^2(x)}} \right]$$

Formulate this BVP in standard form. As in the Sturm–Liouville example, you can introduce a new variable $y_3(x)$, a first-order ODE, and a boundary condition to deal with the integral term in the definition of $H$. The trick to dealing with $H$ is to let it be a new variable $y_4(x)$. It is a constant, so this new variable satisfies the first-order differential equation $y'_4 = 0$. It is given the correct constant value by the boundary condition resulting from the definition of $H$:

$$y_4(1) = \frac{1}{\alpha^2} [1 - (1 - \alpha^2) y_3(1)]$$

### ■ EXERCISE 1.10

This exercise is based on material from the textbook *Continuous and Discrete Signals and Systems* (Soliman & Srinath 1998). A linear, time-invariant (LTI) system is described by a single linear, constant-coefficient ODE of the form

$$y^{(N)}(t) + \sum_{i=0}^{N-1} a_i y^{(i)}(t) = \sum_{i=0}^{N} b_i x^{(i)}(t) \tag{1.19}$$

Here $x(t)$ is a given signal and $y(t)$ is the response of the system. A simulation diagram is a representation of the system using only amplifiers, summers, and integrators. This might be described in many ways, but there are two canonical forms. A state-variable description of a system has some advantages, one being that it is a first-order system of ODEs that is convenient for numerical solution. The two canonical forms for simulation diagrams lead directly to two state-variable descriptions. Let $v(t) = (v_1(t), v_2(t), \ldots, v_N(t))^T$ be a vector of state variables. The description corresponding to the first canonical form is

$$v'(t) = \begin{pmatrix} -a_{N-1} & 1 & 0 & \cdots & 0 \\ -a_{N-2} & 0 & 1 & \cdots & 0 \\ \vdots & \vdots & \vdots & \vdots & \vdots \\ -a_1 & 0 & 0 & \cdots & 1 \\ -a_0 & 0 & 0 & \cdots & 0 \end{pmatrix} v(t) + \begin{pmatrix} b_{N-1} - a_{N-1}b_N \\ b_{N-2} - a_{N-2}b_N \\ \vdots \\ b_1 - a_1 b_N \\ b_0 - a_0 b_N \end{pmatrix} x(t)$$

The output $y(t)$ is obtained from the equation

$$y(t) = (1, 0, \ldots, 0)^\mathrm{T} v(t) + b_N x(t)$$

Show directly that you can solve the ODE (1.19) by solving this system of first-order ODEs. Keep in mind that all the coefficients are constant. *Hint:* Using the identity

$$y(t) = v_1(t) + b_N x(t)$$

rewrite the equations so that, for $i < N$,

$$v_i'(t) = (b_{N-i} x(t) - a_{N-i} y(t)) + v_{i+1}(t)$$

Differentiate the equation for $v_1'(t)$ and use the equation for $v_2'(t)$ to obtain an equation for $v_1''(t)$ involving $v_3(t)$. Repeat until you have an equation for $v_1^{(N)}(t)$, equate it to $(y(t) - b_N x(t))^{(N)}$, and compare the result to the ODE (1.19).

The description corresponding to the second canonical form is

$$v'(t) = \begin{pmatrix} 0 & 1 & 0 & \cdots & 0 \\ 0 & 0 & 1 & \cdots & 0 \\ \vdots & \vdots & \vdots & \vdots & \vdots \\ 0 & 0 & 0 & \cdots & 1 \\ -a_0 & -a_1 & -a_2 & \cdots & -a_{N-1} \end{pmatrix} v(t) + \begin{pmatrix} 0 \\ 0 \\ \vdots \\ 0 \\ 1 \end{pmatrix} x(t)$$

Obtaining the output is more complicated for this form. The formula is

$$y(t) = [(b_0 - a_0 b_N), (b_1 - a_1 b_N), \ldots, (b_{N-1} - a_{N-1} b_N)]^\mathrm{T} v(t) + b_N x(t)$$

Show directly that you can solve the ODE (1.19) by solving this system of first-order ODEs. *Hint:* Define the function $w(t)$ as the solution of the ODE

$$w^{(N)}(t) + \sum_{j=0}^{N-1} a_j w^{(j)}(t) = x(t)$$

and then show by substitution that the function

$$y(t) = \sum_{i=0}^{N} b_i w^{(i)}(t)$$

satisfies the ODE (1.19). Finally, obtain a set of first-order ODEs for the function $w(t)$ in the usual way.

It is striking that the derivatives $x^{(i)}(t)$ do not appear in either of the two canonical systems. Show that they play a role when you want to find a set of initial conditions $v_i(0)$ that corresponds to a set of initial conditions for $y^{(i)}(0)$ and $x^{(i)}(0)$ in the original variables.

# 1.4 Control of the Error

ODE solvers ask how much accuracy you want because the more you want, the more the computation will cost. The MATLAB solvers have error tolerances in the form of a scalar relative error tolerance $re$ and a vector of absolute error tolerances $ae$. The solvers produce vectors $y_n = (y_{n,i})$ that approximate the solution $y(t_n) = (y_i(t_n))$ on the mesh (1.9). Stated superficially, at each point in the mesh they aim to produce an approximation that satisfies

$$|y_i(t_n) - y_{n,i}| \leq re|y_i(t_n)| + ae_i \tag{1.20}$$

for each component of the solution. Variants of this kind of control are seen in all the popular IVP solvers. For the convenience of users, the MATLAB solvers interpret a scalar absolute error tolerance as applying to all components of the solution. Also for convenience, default error tolerances are supplied. They are $10^{-3}$ for the relative error tolerance and a scalar $10^{-6}$ for the absolute error tolerance. The default relative error tolerance has this value because solutions are usually interpreted graphically in MATLAB. A relative error tolerance of $10^{-5}$ is more typical of general scientific computing.

For a code with a vector of relative error tolerances RTOL and a vector of absolute error tolerances ATOL, Brenan, Campbell, & Petzold (1996, p. 131) state:

> We cannot emphasize strongly enough the importance of carefully selecting these tolerances to accurately reflect the scale of the problem. In particular, for problems whose solution components are scaled very differently from each other, it is advisable to provide the code with vector valued tolerances. For users who are not sure how to set the tolerances RTOL and ATOL, we recommend starting with the following rule of thumb. Let $m$ be the number of significant digits required for solution component $y_i$. Set $RTOL_i = 10^{-(m+1)}$. Set $ATOL_i$ to the value at which $|y_i|$ is essentially insignificant.

Because we agree about the importance of selecting appropriate error tolerances, we have devoted this section to a discussion of the issues. This discussion will help you understand the rule of thumb.

The inequality (1.20) defines a *mixed* error control. If all the values $ae_i = 0$, it corresponds to a pure *relative* error control; if the value $re = 0$, it corresponds to a pure *absolute* error control. The pure error controls expose more clearly the roles of the two kinds of tolerances and the difficulties associated with them. First suppose that we use a pure relative error control. It requires that

$$\left| \frac{y_i(t_n) - y_{n,i}}{y_i(t_n)} \right| \le re$$

for each solution component. There are two serious difficulties. One is that a pure relative error control is not appropriate if the solution might vanish. The formal difficulty is that the denominator $y_i(t_n)$ might vanish. However, we are attempting to control the error in a function, so the more fundamental question is: What should we mean by relative error if $y_i(t)$ might vanish at some isolated point $t = t^*$? The solvers commonly compare the error to some measure of the size of $y_i(t)$ near $t_n$ rather than just the value $|y_i(t_n)|$ of (1.20). This is a reasonable and effective approach, but it does not deal with a component $y_i(t)$ that is zero throughout an interval about $t_n$. Solvers must therefore recognize the possibility that a relative error control is not well-defined, even in some extended sense, and terminate the integration with a message should this occur. You can avoid the difficulty by specifying a nonzero absolute error tolerance in a mixed error test. For robustness some solvers, including those of MATLAB, require that absolute error tolerances be positive.

Before taking up the other difficulty, we need to make some comments about computer arithmetic. Programming languages like Fortran 77 and C include both single and double precision arithmetic. Typically this corresponds to about 7 and 16 decimal digits, respectively. MATLAB has only one precision, typically double precision. Experience says that, when solving IVPs numerically, it is generally best to use double precision. The floating point representation of a number is accurate only to a unit roundoff, which is determined by the working precision. In MATLAB it is called eps and for a PC it is typically $2.2204 \cdot 10^{-16}$, corresponding to double precision in the IEEE-754 definition of computer arithmetic that is used almost universally on today's computers. Throughout this book we assume that the unit roundoff is about this size when we speak of computations in MATLAB.

A relative error tolerance specifies roughly how many correct digits you want in an answer. It makes no sense to ask for an answer more accurate than the floating point representation of the true solution – that is, it is not meaningful to specify a value $re$ smaller than a unit roundoff. Of course, a tolerance that is close to a unit roundoff is usually also too small because finite precision arithmetic affects the computation and hence the accuracy that a numerical method can deliver. For this reason the MATLAB solvers require that $re$ be larger than a smallish multiple of eps, with the multiple depending on the particular solver. You might expect that a code would fail in some dramatic way if you ask for an impossible accuracy. Unfortunately, that is generally not the case. If you experiment with a code that does not check then you are likely to find that, as you decrease the tolerances past the point where you are requesting an impossible accuracy: the cost of the integration increases rapidly; the results are increasingly *less* accurate; and there is no indication from the solver that it is having trouble, other than the increase in cost.

Now we turn to a pure absolute error control. It requires that

$$|y_i(t_n) - y_{n,i}| \leq ae_i$$

for each solution component. The main difficulty with an absolute error control is that you must make a judgment about the likely sizes of solution components, and you can get into trouble if you are badly wrong. One possibility is that a solution component is much larger in magnitude than expected. A little manipulation of the absolute error control inequality leads to

$$\left| \frac{y_i(t_n) - y_{n,i}}{y_i(t_n)} \right| \leq \frac{ae_i}{|y_i(t_n)|}$$

This makes clear that a pure absolute error tolerance of $ae_i$ on $y_i(t)$ corresponds to a relative error tolerance of $ae_i/|y_i(t_n)|$ on this component. If $|y_i(t_n)|$ is sufficiently large, then specifying an absolute error tolerance that seems unremarkable can correspond to asking for an answer that is more accurate in a relative sense than a unit roundoff. As we have just seen, that is an impossible accuracy request. The situation can be avoided by specifying a nonzero relative error tolerance and thus a mixed error control. Again for the sake of robustness, the MATLAB solvers do this by requiring that the relative error tolerance be greater than a few units of roundoff.

The other situation that concerns us with pure absolute error control is when a solution component is much smaller than its absolute error tolerance. First we must understand what the error control means for such a component. If (say) $|y_i(t_n)| < 0.5ae_i$, then *any* approximation $y_{n,i}$ for which $|y_{n,i}| < 0.5ae_i$ will pass the error test. Accordingly, an acceptable approximation may have *no* correct digits. You might think that you always need some accuracy, but for many mathematical models of physical processes there are quantities that have negligible effects when they fall below certain thresholds and are then no longer interesting. The danger is that one of these quantities might later grow to the point that it must again be taken into account. If a solution component is rather smaller in magnitude than its absolute error tolerance and if you require some accuracy in this component, you will need to adjust the tolerance and solve the problem again. It is an interesting and useful fact that you may very well compute some correct digits in a "small" component even though you did not require it by means of its error tolerance. One reason is that the solver may have computed this component with some accuracy in order to achieve the accuracy specified for a component that depends on it. Another reason is that the solver selects a step size small enough to deal with the solution component that is most difficult to approximate to within the accuracy specified. Generally this step size is smaller than necessary for other components, so they are computed more accurately than required.

The first example of Lapidus, Aiken, & Liu (1973) is illustrative. Proton transfer in a hydrogen–hydrogen bond is described by the system of ODEs

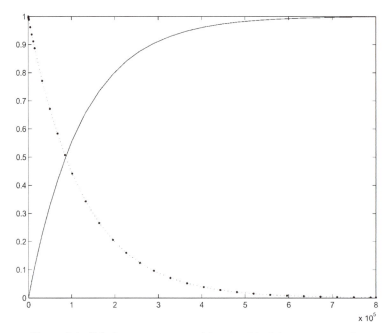

Figure 1.6: Solution components $x_1(t)$ and $x_2(t)$ of the proton transfer problem.

$$\begin{aligned}
x_1' &= -k_1 x_1 + k_2 y \\
x_2' &= -k_4 x_2 + k_3 y \\
y' &= k_1 x_1 + k_4 x_2 - (k_1 + k_3) y
\end{aligned} \tag{1.21}$$

to be solved with initial values

$$x_1(0) = 0, \quad x_2(0) = 1, \quad y(0) = 0$$

on the interval $0 \leq t \leq 8 \cdot 10^5$. The coefficients here are

$$k_1 = 8.4303270 \cdot 10^{-10}, \quad k_2 = 2.9002673 \cdot 10^{11},$$
$$k_3 = 2.4603642 \cdot 10^{10}, \quad k_4 = 8.7600580 \cdot 10^{-6}$$

This is an example of a *stiff* problem. We solved it easily with the MATLAB IVP solver `ode15s` using default error tolerances, but we found that the quickly reacting intermediate component $y(t)$ is very much smaller than the default absolute error tolerance of $10^{-6}$. Despite this, it was computed accurately enough to give a general idea of its size. Once we recognized how small it is, we reduced the absolute error tolerance to $10^{-20}$ and obtained the solutions displayed in Figures 1.6 and 1.7. It is easy and natural in exploratory computations with the MATLAB ODE solvers to display all the solution components on one plot. If some components are invisible then you might want to determine

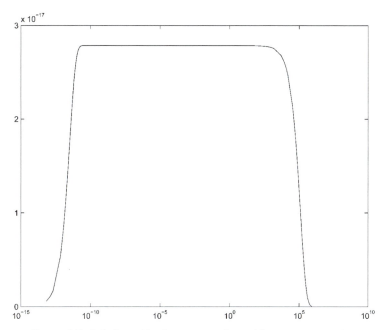

Figure 1.7: Solution $y(t)$ of proton transfer problem, `semilogx` plot.

the maximum magnitudes of the solution components – both to identify components for plotting separately on different scales and for choosing tolerances for another, more accurate computation.

Often in modeling chemical reactions, concentrations that have dropped below a certain threshold have negligible effects and so are of no physical interest. Then it is natural to specify absolute error tolerances of about the sizes of these thresholds. The concentrations $y_i(t)$ are positive, but when tracking a component $y_i(t)$ that decays to zero a solver might generate a "small" solution component $y_{n,i} < 0$. As we have seen, the error control permits this and it sometimes happens. A small negative approximation to a concentration may just be an annoyance, but some models are not stable in these circumstances and the computation blows up. It is ironic that a quantity so small that it is unimportant physically can destroy the numerical solution. An IVP popularized by Robertson (1966) as a test problem for solvers intended for stiff IVPs provides a concrete example. A chemical reaction is described by the system of ODEs

$$
\begin{aligned}
y_1' &= -0.04y_1 + 10^4 y_2 y_3 \\
y_2' &= 0.04y_1 - 10^4 y_2 y_3 - 3 \cdot 10^7 y_2^2 \\
y_3' &= 3 \cdot 10^7 y_2^2
\end{aligned}
\tag{1.22}
$$

with initial conditions

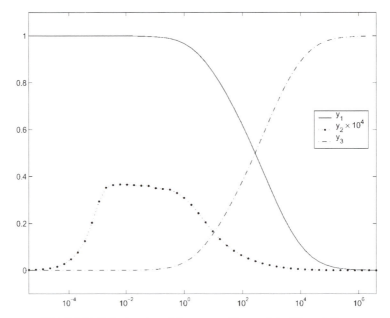

Figure 1.8: Robertson's problem; a `semilogx` plot of the solution.

$$\begin{pmatrix} y_1(0) \\ y_2(0) \\ y_3(0) \end{pmatrix} = \begin{pmatrix} 1 \\ 0 \\ 0 \end{pmatrix}$$

It is not difficult to show that, for all times $t > 0$, the solution components are nonnegative and sum to 1. This is an example of a linear conservation law that we will discuss in some detail in the next section.

The `hb1ode` demonstration program of MATLAB integrates this problem with `ode15s` from time $t = 0$ to near steady state at $t = 4 \cdot 10^6$. A small modification of its output resulted in Figure 1.8. Hindmarsh & Byrne (1976) use this problem to illustrate the performance of their code `EPISODE` for solving stiff IVPs. With a moderately stringent absolute error tolerance of $10^{-6}$, they find that if they continue the integration then a small non-physical negative concentration is computed that begins growing rapidly in magnitude. Soon the numerical solution is completely unacceptable. A portion of one of their tables of results is given in Table 1.1. We emphasize that the unsatisfactory performance is a consequence of the problem and what is asked of the solver; something similar happens when other solvers are used, including `ode15s`. For more details about this example see Hindmarsh & Byrne (1976) and Shampine (1994). Different but related problems are considered in Exercises 1.12 and 1.13.

We have seen that you cannot ask for too much accuracy in a relative sense. We take this opportunity to advise you not to ask for too little. This is a temptation because the

Table 1.1: *Robertson's problem; steady-state solution computed using* EPISODE.

| $t$ | $y_1$ | $y_2$ | $y_3$ |
|---|---|---|---|
| 4e5 | 4.9394e−03 | 1.9854e−08 | 9.9506e−01 |
| 4e7 | 3.2146e−05 | 1.2859e−10 | 9.9997e−01 |
| 4e9 | −1.8616e+06 | −4.0000e−06 | 1.8616e+06 |

more accuracy you want, the more the computation will cost. It is especially tempting when the data of a problem is known only to a digit or two. (We have solved IVPs for which even the order of magnitude of measured data was in doubt.) Nevertheless, asking for too little accuracy is both dangerous and pointless. The basic algorithms are valid only when the step sizes are sufficiently small. If you do not ask for enough accuracy, a solver might choose step sizes that are too large for reliable results. A quality solver may recognize that it must use smaller step sizes for reliability and in effect reduce the error tolerances that you specify. As explained in Section 1.2, the solvers control local errors and only indirectly control the error in the solution $y(t)$. They maintain these local errors somewhat smaller than the tolerances. How much smaller is "tuned" for the solver so that, for typical IVPs, the error in $y(t)$ is smaller than (or comparable to) the tolerances specified. If your IVP is somewhat unstable or you expect the solution to oscillate often in the interval of interest, then you should be cautious about asking for too little accuracy because you might well be disappointed in the accuracy that you get. That is the least of your worries: You might compute a solution that is not physically realistic, or one that is physically realistic but incorrect, or the computation might fail entirely. In considering this it is important to appreciate that the solver is doing exactly what you tell it to do – namely, to control the local error so that it is no larger than the specified tolerances. Unsatisfactory results are usually a consequence of the instability of the IVP, not of the solver. Figure 1.9 of Section 1.5 shows what can happen. The dotted curve is an orbit of one body about another that was computed with default error tolerances. These default error tolerances were intended to be satisfactory for plotting the solutions of typical problems, but in this instance the orbit is not even qualitatively correct. Displayed as a solid curve is the same orbit computed with more stringent error tolerances. It is qualitatively correct. Clearly it is important not to ask for too little accuracy when solving this problem.

When solving a newly formulated IVP, it may be necessary to experiment with the choice of error tolerances. To do this, you may need to inspect solutions to verify that you are using an appropriate error control. You may also want to try reducing the error tolerances to verify by consistency that you are asking for sufficient accuracy to reflect the qualitative behavior of the solution.

■ **EXERCISE 1.11**

To simplify their user interface, some codes ask for a single error tolerance $\tau$. For example, DVERK (Hull, Enright, & Jackson 1975) requires that, at each step,

$$|y_i(t_n) - y_{n,i}| \leq \tau \max(1, |y_i(t_n)|)$$

and MIRKDC (Enright & Muir 1996) requires the equivalent of

$$|y_i(t_n) - y_{n,i}| \leq \tau(1 + |y_i(t_n)|)$$

Argue that these are roughly equivalent to the error control (1.20) with $re = \tau$ and $ae_i = \tau$ for each $i$. People sometimes get into trouble with this kind of error control because they do not realize that they are specifying an absolute error tolerance that is not appropriate for the problem they are solving.

■ **EXERCISE 1.12**

The solution of

$$y' = f(t, y) = \sqrt{1 - y^2}, \quad y(0) = 0$$

is $\sin(t)$. When computing this solution numerically, why should you expect to get into trouble as you approach $t = 0.5\pi$? There are two kinds of difficulties, one involving the error control and one involving uniqueness.

■ **EXERCISE 1.13**

If you solve the IVP

$$y' = \left(\frac{2\ln(y) + 8}{t} - 5\right)y, \quad y(1) = 1$$

with a code written in MATLAB, you might compute approximations to $y(t)$ that are complex-valued for "large" $t$. Codes in other computing environments might fail outright. What is going on? To answer this question it is helpful to know that the solution is

$$y(t) = e^{-t^2 + 5t - 4}$$

# 1.5 Qualitative Properties

We have seen several examples of solutions with certain qualitative properties that are implied by the ODEs. It is commonly assumed that numerical solutions inherit these properties, but with one major exception they do *not*. The best we can say for standard methods is that the numerical solutions have approximately the same behavior as the analytical solutions. There are ways of making standard methods do better in this regard and there are methods that preserve certain qualitative properties, but we do not pursue such specialized aspects of solving ODEs in this book. For further information about these matters you might turn to Sanz-Serna & Calvo (1994), Shampine (1986), Stuart & Humphries (1996), and the references therein.

We begin our discussion of qualitative properties with one that *is* inherited by virtually all standard methods. If there is a constant (column) vector $c$ such that $c^{\mathrm{T}} f(t, y) \equiv 0$, then the solution of the ODE system

$$y' = f(t, y), \quad y(a) = A$$

satisfies the *linear conservation law*,

$$c^{\mathrm{T}} y(t) \equiv c^{\mathrm{T}} A$$

This follows on observing that

$$\frac{d}{dt}(c^{\mathrm{T}} y(t)) = c^{\mathrm{T}} y'(t) = c^{\mathrm{T}} f(t, y(t)) \equiv 0$$

and hence $c^{\mathrm{T}} y(t)$ is constant. Linear conservation laws express physical laws such as conservation of mass and charge balance. The hydrogen–hydrogen bond problem (1.21) and Robertson's problem (1.22) are examples. With the initial values specified, the solutions of both these problems have components that sum to 1. As it turns out (Shampine 1998), all the standard numerical methods for IVPs preserve all linear conservation laws. For example, if the components of the solution sum to 1, then so do the components of the numerical approximation (to within roundoff errors). The fact that the numerical solution satisfies one or more conservation laws does not mean that it is accurate – even the terrible numerical solution of Robertson's problem found in Table 1.1 has components that sum to 1. On the other hand, if a linear conservation law is not satisfied by the numerical solution to roundoff level, then there is a bug in the program that produced it or the computations were overwhelmed by the effects of finite precision arithmetic. We turn now to properties that are not preserved by standard methods.

In Section 1.2 we found that solutions of the pendulum equation (1.6) have a constant energy. Generally the numerical solutions computed with standard software have an energy that is only approximately constant. To see that it is at least approximately constant, suppose that the equation is written as a first-order system. Further suppose that, at time $t_n$, the solver produces approximations

$$y_{n,1} = \theta(t_n) + e_1, \quad y_{n,2} = \theta'(t_n) + e_2$$

with small errors $e_1$ and $e_2$. By linearization we approximate the energy of the numerical solution as

$$0.5(y_{n,2})^2 - \cos(y_{n,1}) = 0.5(\theta'(t_n) + e_2)^2 - \cos(\theta(t_n) + e_1)$$
$$\approx E + \theta'(t_n)e_2 + \sin(\theta(t_n))e_1$$

This tells us that the error in the energy is comparable to the errors in the solution components; hence the energy is approximately constant. Often this is satisfactory. However, the long-term qualitative behavior of solutions may depend on the energy and it may be important to conserve energy. One way to do this is simply to solve the equations very accurately using a standard code. This may be satisfactory for short to medium time scales. Alternatively, there are codes based on standard methods that optionally perturb the numerical solution so that it satisfies specified nonlinear conservation laws. There are also codes based on methods that automatically conserve certain physically important quantities, usually energy and/or angular momentum. Whether it is more efficient to use one of these specialized codes or to ask for more accuracy from a standard code is a matter for experimentation. In many cases conservation of a nonlinear conservation law may only be achieved at a high cost or at the expense of accuracy in the solution.

Solutions of the two-body problem (1.17) satisfy two nonlinear conservation laws. The energy

$$\frac{x'(t)^2 + y'(t)^2}{2} - \frac{1}{r(t)}$$

(where the distance $r(t) = \sqrt{x(t)^2 + y(t)^2}$) and the angular momentum

$$x(t)y'(t) - y(t)x'(t)$$

are constant. Figure 1.9 shows the solution of the ODE system (1.17) with initial conditions (1.18) when the eccentricity $e = 0.9$. The path of the moving body displayed as a solid curve was computed with moderately stringent tolerances. The other path was computed with default error tolerances. The fixed body at the origin is shown as an asterisk. For this problem and particular choice of integrator, the energy of the numerical solution computed with default error tolerances decreases steadily from $-0.5000$ to $-0.7874$ while the angular momentum decreases from $0.4359$ to $0.3992$. A steady loss of energy in the physical problem corresponds to the moving body spiraling in to the fixed body. What happens in a numerical computation will depend on the method used and the details of the problem. The point, however, is that the numerical solution satisfies the conservation laws only approximately. Over a time interval sufficiently long, the numerical solution might have a behavior that is qualitatively different from the mathematical solution. Because this particular integrator is losing energy steadily for this particular problem, the effect is pronounced. On the other hand, when we tell the integrator to compute a more accurate answer by specifying smaller error tolerances, we compute a solution on $[0, 20]$ that has the expected behavior. There has been a small loss of energy by time $t = 20$ in this integration, but it is too small for the effect on the computed solution to be visible in the plot.

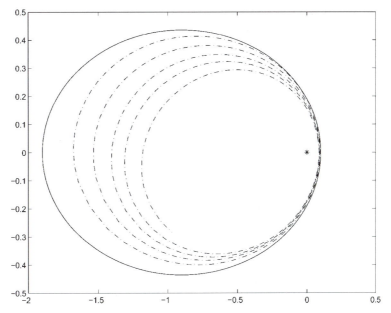

Figure 1.9: Two-body problem for $e = 0.9$ and $0 \leq t \leq 20$.

There are methods that preserve (at least approximately) certain qualitative properties of IVPs over extended integrations. For example, for ODEs that define symplectic or time-reversible maps, it is possible to construct numerical methods with the corresponding property. These methods bound the error in the Hamiltonian energy and, in some cases, conserve angular momentum; see Sanz-Serna & Calvo (1994) or Stuart & Humphries (1996) for details. Of course, these desirable properties come at a price. The additional constraints placed on the methods to achieve a special property such as symplecticness potentially reduce the accuracy that can be achieved in the computation of the solution at a given cost.

■ **EXERCISE 1.14**

The differential equations

$$y_1' = -y_1$$
$$y_k' = (k-1)y_{k-1} - ky_k \quad \text{for } k = 2, 3, \ldots, 9$$
$$y_{10}' = 9y_9$$

describe the evolution of a chemical reaction. Show that this system of ODEs satisfies a linear conservation law. Specifically, show that the sum of the solution components is constant.

■ **EXERCISE 1.15**

Volterra's model of predator–prey interaction can be formulated as

$$x' = a(x - xy)$$
$$y' = -c(x - xy)$$

- Show that solutions of this system of ODEs satisfy the nonlinear conservation law

$$G(t, x, y) = x^{-c}y^{-a}e^{cx+ay} = \text{constant}$$

- Write a MATLAB program to integrate the differential equations with Euler's method and constant step size $h$. Using parameter values $a = 2$ and $c = 1$ and initial values $x(0) = 1$ and $y(0) = 3$, integrate the IVP for $0 \leq t \leq 10$. Plot the solution in the phase plane; that is, plot $(x(t), y(t))$. Also, calculate and plot the conserved quantity $G(t, x(t), y(t))$. The theory says that $G$ is constant and the solution is periodic, hence the curve plotted in the phase plane is closed. Experiment with the step size $h$ to find a value for which $G$ is approximately constant and the curve you compute appears to be closed. After you have learned to use the MATLAB IVP solvers in the next chapter, you may want to revisit this problem and solve it with ode45 instead of Euler's method.

# Chapter 2

## Initial Value Problems

## 2.1  Introduction

In this chapter we study the solution of initial value problems for ordinary differential equations. Because ODEs arise in diverse forms, it is convenient for both theory and practice to write them in a standard form. It was shown in Chapter 1 how to prepare ODEs as a system of first-order equations that in (column) vector notation has the form

$$y' = f(t, y) \tag{2.1}$$

With one exception, it is assumed throughout this chapter that the ODEs have this form. Because the MATLAB IVP solvers accept problems of the form $M(t, y)y' = f(t, y)$, it is discussed briefly in Section 2.3.2. In either case it is assumed that the ODEs are defined on a finite interval $a \le t \le b$ and that the initial values are provided as a vector

$$y(a) = A \tag{2.2}$$

The popular numerical methods for IVPs start with $y_0 = A = y(a)$ and then successively compute approximations $y_n \approx y(t_n)$ on a mesh $a = t_0 < t_1 < \cdots < t_N = b$. On reaching $t_n$, the basic methods are distinguished by whether or not they use previously computed quantities such as $y_{n-1}, y_{n-2}, \ldots$. If they do, they are called *methods with memory* and otherwise, *one-step methods*. IVPs are categorized as nonstiff and stiff. It is hard to define stiffness, but its symptoms are easy to recognize. Unfortunately, the distinction between stiff and nonstiff IVPs can be very important when choosing a method. The MATLAB IVP solvers implement a variety of methods, but the documentation recommends that you first try `ode45`, a code based on a pair of one-step *explicit Runge–Kutta formulas*. If you suspect that the problem is stiff or if `ode45` should prove unsatisfactory, it is recommended that you try `ode15s`, a code based on the *backward differentiation formulas* (BDFs).

These two types of methods are among the most widely used in general scientific computing, so we focus on them in our discussion of numerical methods. We do take up other methods to provide some perspective; in particular, we discuss the *Adams methods* that are implemented in `ode113`. They are often preferred over explicit Runge–Kutta methods when solving nonstiff problems in general scientific computing. The last part of this chapter is a tutorial that shows how to solve IVPs with the programs of MATLAB. You can read it and solve interesting problems in parallel with your reading about the theory of the various methods implemented in the programs.

## 2.2  Numerical Methods for IVPs

We focus on two kinds of methods for solving IVPs, the ones used by `ode45` and `ode15s`. They are explicit Runge–Kutta formulas for nonstiff IVPs and backward differentiation formulas for stiff IVPs, respectively. These methods are unquestionably among the most effective and widely used. Comments are made about other methods where this helps put the developments in perspective. Although our discussion of the methods is brief, it does identify the most important issues for solving IVPs in practice.

Numerical solution of the IVP (2.1), (2.2) on the interval $a \leq t \leq b$ proceeds in steps. Starting with the initial value $y_0 = A$, values $y_n \approx y(t_n)$ are computed successively on a mesh

$$a = t_0 < t_1 < \cdots < t_N = b$$

The computation of $y_{n+1}$ is often described as taking a step of size $h_n = t_{n+1} - t_n$ from $t_n$. For brevity we generally write $h = h_n$ in discussing the step from $t_n$. On reaching $(t_n, y_n)$, the *local solution* $u(t)$ is defined as the solution of

$$u' = f(t, u), \quad u(t_n) = y_n$$

A standard result from the theory of ODEs states that if $v(t)$ and $w(t)$ are solutions of (2.1) and if $f(t, y)$ satisfies a Lipschitz condition with constant $L$, then for $\alpha < \beta$ we have

$$\|v(\beta) - w(\beta)\| \leq \|v(\alpha) - w(\alpha)\| e^{L(\beta - \alpha)}$$

In the *classical situation* that $L(b - a)$ is of modest size, this result tells us that the IVP (2.1), (2.2) is moderately stable. This is only a sufficient condition. Indeed, *stiff* problems are (very) stable, yet $L(b - a) \gg 1$. Without doing some computation, it is not easy to recognize that a stable IVP is stiff. There are two essential properties that will help you with this: A stiff problem is very stable in the sense that some solutions of the ODE starting near the solution of interest converge to it very rapidly ("very rapidly" here means that the solutions converge over a distance that is small compared to $b - a$, the length of the interval of integration). This property implies that some solutions change very rapidly, but the second property is that the solution of interest is slowly varying.

The basic numerical methods approximate the solution only on a mesh, but in some codes – including all of the MATLAB solvers – they are supplemented with (inexpensive) methods for approximating the solution between mesh points. The BDFs are based on polynomial interpolation and so give rise immediately to a continuous piecewise-polynomial function $S(t)$ that approximates $y(t)$ everywhere in $[a, b]$. There is no natural polynomial interpolant for explicit Runge–Kutta methods, which is why such interpolants are a relatively new development. A method that approximates $y(t)$ on each step $[t_n, t_{n+1}]$ by a polynomial that interpolates the approximate solution at the end points of the interval is called a *continuous extension* of the Runge–Kutta formula. We'll use the term more generally to refer to a piecewise-polynomial approximate solution $S(t)$ defined in this way on all of $[a, b]$.

## 2.2.1  One-Step Methods

One-step methods use only data gathered in the current step. MATLAB includes solvers based on a number of different kinds of one-step methods, but we concentrate on one kind: explicit Runge–Kutta methods. In the course of our study of these explicit methods, we'll also develop some implicit methods. They are widely used for solving BVPs, so we return to them in Chapter 3.

It is illuminating first to take up the special case of *quadrature*:

$$y' = f(t), \quad y(a) = A \tag{2.3}$$

Certainly we must be able to deal with these simpler problems and, because they are simpler, it is easier to explain the methods. The local solution at $t_n$ satisfies

$$u' = f(t), \quad u(t_n) = y_n$$

so

$$u(t_n + h) = y_n + \int_{t_n}^{t_n+h} f(x)\, dx$$

Computing $y_{n+1} \approx u(t_n + h)$ amounts to approximating numerically a definite integral.

A basic tactic in numerical analysis is this: If you cannot do what you want with a function $f(x)$, approximate it with an interpolating polynomial $P(x)$ and use the polynomial instead. The popular formulas for approximating definite integrals can be derived in this way. For instance, if we approximate the function $f(x)$ on the interval $[t_n, t_n + h]$ by interpolating it with the constant polynomial $P(x) = f(t_n)$, we can integrate the polynomial to obtain

$$\int_{t_n}^{t_n+h} f(x)\, dx \approx \int_{t_n}^{t_n+h} P(x)\, dx = h f(t_n)$$

Or, if we interpolate at the other end of the interval,

$$\int_{t_n}^{t_n+h} f(x)\, dx \approx \int_{t_n}^{t_n+h} P(x)\, dx = h f(t_{n+1})$$

Similarly, if we use a linear polynomial that interpolates $f(x)$ at both ends of the interval,

$$P(x) = \left(\frac{(t_n+h)-x}{h}\right) f(t_n) + \left(\frac{x-t_n}{h}\right) f(t_n+h)$$

then by integrating we obtain the approximation

$$\int_{t_n}^{t_n+h} f(x)\, dx \approx \frac{h}{2}[f(t_n) + f(t_n+h)]$$

Geometrically, the first two schemes approximate the integral by the area of a rectangle. The third is known as the trapezoidal rule because it approximates the integral by the area of a trapezoid.

We can deduce the accuracy of these approximations for a smooth function $f(x)$ by a standard result from polynomial interpolation theory (Shampine, Allen, & Pruess 1997): If $P(x)$ is the unique polynomial of degree less than $s$ that interpolates a smooth function $f(x)$ at $s$ distinct nodes $t_{n,j} = t_n + \alpha_j h$ in the interval $[t_n, t_n + h]$,

$$P(t_{n,j}) = f(t_{n,j}), \quad j = 1, 2, \ldots, s$$

then

$$P(x) = \sum_{j=1}^{s} f_{n,j} \prod_{i=1, i \neq j}^{s} \frac{x - t_{n,i}}{t_{n,j} - t_{n,i}}$$

and, for each point $x$ in the interval $[t_n, t_n + h]$, there is a point $\xi$ in the same interval for which

$$f(x) - P(x) = \frac{f^{(s+1)}(\xi)}{(s+1)!} \prod_{j=1}^{s} (x - t_{n,j})$$

For a function $f(x)$ that is sufficiently smooth, the derivative appearing in this expression is bounded in magnitude on the interval. It is then easy to see that there is a constant $C$ such that

$$|f(x) - P(x)| \leq C h^s$$

for all $x \in [t_n, t_n + h]$. A standard notation and terminology is used when we focus our attention on the behavior with respect to the step size $h$, ignoring the value of the constant. We write

$$f(x) - P(x) = O(h^s)$$

or, equivalently, $f(x) = P(x) + O(h^s)$. We say that the difference between $f(x)$ and $P(x)$ is "big oh of $h$ to the $s$" or, for short, that the difference is of *order s*. With the theorem about the accuracy of polynomial interpolation, it is easy to see that

$$\int_{t_n}^{t_n+h} f(x)\, dx = \int_{t_n}^{t_n+h} P(x)\, dx + O(h^{s+1})$$

Applying these results to the rectangle approximations to the integral gives the formula

$$y_{n+1} = y_n + hf(t_n) \tag{2.4}$$

for which the local error

$$u(t_n + h) - y_{n+1} = \int_{t_n}^{t_n+h} f(x)\, dx - hf(t_n) = O(h^2)$$

and

$$y_{n+1} = y_n + hf(t_{n+1}) \tag{2.5}$$

which has the same order of accuracy. The trapezoidal rule

$$y_{n+1} = y_n + h\left[\tfrac{1}{2}f(t_n) + \tfrac{1}{2}f(t_n + h)\right] \tag{2.6}$$

has local error $u(t_n + h) - y_{n+1} = O(h^3)$. More generally, polynomial interpolation at $s$ nodes leads to an *interpolatory quadrature formula* of the form

$$\int_{t_n}^{t_n+h} f(x)\, dx = h\sum_{j=1}^{s} A_j f(t_n + \alpha_j h) + O(h^{p+1})$$

and to a numerical method

$$y_{n+1} = y_n + h\sum_{j=1}^{s} A_j f(t_n + \alpha_j h) \tag{2.7}$$

for which the local error is $O(h^{p+1})$. Interpolation theory assures us that the order $p \geq s$, but an important fact in the practical approximation of integrals is that, for some choices of nodes, the order $p > s$. For instance, the midpoint rule that comes from a constant polynomial that interpolates $f(x)$ at $t_n + 0.5h$ has order $p = 2$ instead of the value $p = 1$ that we might expect on general grounds.

We have obtained specific formulas of the form (2.7) by integrating an interpolating polynomial. Let us now start with a formula of this general form and ask how we might choose the coefficients $A_j$ to find an accurate formula. To determine how accurate the formula is, we'll expand both $u(t_n + h)$ and $y_{n+1}$ in Taylor series about $t_n$ and see how many terms agree. Using the IVP satisfied by the local solution, we find that

$$u(t_n + h) = u(t_n) + \sum_{k=1}^{p} h^k \frac{u^{(k)}(t_n)}{k!} + O(h^{p+1})$$

$$= y_n + \sum_{k=1}^{p} h^k \frac{f^{(k-1)}(t_n)}{k!} + O(h^{p+1})$$

Expanding the formula on the right-hand side of (2.7) is a little more complicated. First, using Taylor series we expand

$$f(t_n + \alpha_j h) = \sum_{r=0}^{p-1} (\alpha_j h)^r \frac{f^{(r)}(t_n)}{r!} + O(h^p)$$

and then substitute this result into the formula to obtain

$$y_{n+1} = y_n + h \sum_{j=1}^{s} A_j \left( \sum_{k=1}^{p} (\alpha_j h)^{k-1} \frac{f^{(k-1)}(t_n)}{(k-1)!} \right) + O(h^{p+1})$$

$$= y_n + \sum_{k=1}^{p} h^k \left( \sum_{j=1}^{s} A_j \alpha_j^{k-1} \right) \frac{f^{(k-1)}(t_n)}{(k-1)!} + O(h^{p+1})$$

Comparing the two expansions, we see that $u(t_n + h) = y_{n+1} + O(h^{p+1})$ if and only if

$$\frac{1}{k} = \sum_{j=1}^{s} A_j \alpha_j^{k-1}, \quad k = 1, 2, \ldots, p \tag{2.8}$$

If we use only one node (i.e., $s = 1$) then the first equation of (2.8) requires that $A_1 = 1$. With this value for $A_1$, the second equation is

$$\frac{1}{2} = A_1 \alpha_1 = \alpha_1$$

If $\alpha_1 \neq \frac{1}{2}$, this equation is not satisfied and $u(t_n + h) = y_{n+1} + O(h^2)$. If $\alpha_1 = \frac{1}{2}$, the equation is satisfied and the third equation becomes

$$\frac{1}{3} = 1 \times \left( \frac{1}{2} \right)^2$$

This equation is not satisfied, so $u(t_n + h) = y_{n+1} + O(h^3)$ for this formula. The formula is the midpoint rule that we just met. Exercise 2.1 asks you to verify the order of two other formulas that are based on important quadrature rules.

We begin a discussion of the convergence of these formulas by working out the stability of the ODE. The function $f$ satisfies a Lipschitz condition with $L = 0$, so the general result stated earlier tells us that if $v(t)$ and $w(t)$ are solutions of $y' = f(t)$ and if $\alpha < \beta$, then

$$\|v(\beta) - w(\beta)\| \leq \|v(\alpha) - w(\alpha)\|$$

However, it is perfectly easy to prove a stronger result directly. A solution $v(t)$ of $y' = f(t)$ has the form

$$v(t) = v(\alpha) + \int_\alpha^t f(x)\, dx$$

Evaluating this expression at $t = \beta$ and subtracting a similar expression for $w(t)$ leads to

$$v(\beta) - w(\beta) = v(\alpha) - w(\alpha)$$

From this we see that the local error of a step from $t_n$ moves us to a solution of the ODE that is parallel to the solution through $y_n$.

Let the true error at $t_n$ be

$$e_n = y(t_n) - y_n$$

By choosing $y_0 = A$, we have $e_0 = 0$. In Chapter 1 we studied the propagation of error by writing

$$e_{n+1} = y(t_{n+1}) - y_{n+1} = [u(t_{n+1}) - y_{n+1}] + [y(t_{n+1}) - u(t_{n+1})]$$

The first term on the right is the local error that we assume is bounded in magnitude by $Ch_n^{p+1}$. Using the general bound on stability for ODEs that satisfy Lipschitz conditions, the second term is bounded by

$$|y(t_{n+1}) - u(t_{n+1})| \leq |y(t_n) - y_n| e^{Lh_n}$$

Here we use the definition $u(t_n) = y_n$. Putting these bounds together, we obtain

$$|e_{n+1}| \leq Ch_n^{p+1} + |e_n| e^{Lh_n}$$

The error in this step comes from two sources. One is the local error introduced at each step by the numerical method; the other is amplification of the error from preceding steps due to the stability of the IVP itself. The net effect is particularly easy to understand for quadrature problems because there is no amplification and the local errors just add up in this bound. If we solve a quadrature problem with a constant step size $h = (b - a)/N$, we have a uniform bound

$$|y(t_n) - y_n| = |e_n| \leq nCh^{p+1} \leq (b - a)Ch^p$$

That is, $y_n = y(t_n) + O(h^p)$ for all $n$. For this reason, when the local error is $O(h^{p+1})$, we say that the formula is of order $p$ because that is the order of approximation to $y(t)$. When the step size varies, it is easy to modify this proof to see that if $H$ is the maximum step size then the true (global) error is $O(H^p)$.

## Local Error Estimation

The local error of the result $y_{n+1}$ of a formula of order $p$ is

$$le_n = u(t_n + h) - y_{n+1}$$

If we also apply a formula of order $p + 1$ to compute a result $y_{n+1}^*$ on this step, we can form

$$
\begin{aligned}
est &= y_{n+1}^* - y_{n+1} \\
&= [u(t_n + h) - y_{n+1}] - [u(t_n + h) - y_{n+1}^*] \\
&= le_n + O(h^{p+2})
\end{aligned}
\tag{2.9}
$$

This is a computable estimate of the local error of the lower-order formula because $le_n$ is $O(h^{p+1})$ and so dominates in (2.9) for small enough values of $h$. Put differently, we can estimate the error in $y_{n+1}$ by comparing it to the more accurate approximate solution $y_{n+1}^*$. Generally the most expensive part of taking a step is forming the function values $f(t_n + \alpha_j h)$, so the trick to making local error estimation practical is to find a pair of formulas that share as many of these function evaluations as possible. For the estimate to be any good, the higher-order result $y_{n+1}^*$ must be the more accurate. But if that is so, why would we discard it in favor of using $y_{n+1}$ to advance the integration? Advancing the integration with the more accurate result $y_{n+1}^*$ is called *local extrapolation*. Most of the popular explicit Runge–Kutta codes use local extrapolation; in particular, the MATLAB solvers `ode23` and `ode45` use it. In this way of proceeding, we do not know precisely how small the local error is at each step, but we believe that it is rather smaller than the estimated local error.

Solvers for IVPs control the estimated local error. A local error tolerance $\tau$ is specified and, if the estimated error is too large relative to this tolerance, the step is rejected and another attempt is made with a smaller step size. In our expansion of the local error, we worked out the order of only the first nonzero term. If we carry another term in the expansion, we find that

$$u(t_n + h) - y_{n+1} = h^{p+1}\phi(t_n) + O(h^{p+2}) \tag{2.10}$$

Using this, we can see how to adjust the step size. If we were to try again to take a step from $t_n$ with step size $\sigma h$, the local error would be

$$(\sigma h)^{p+1}\phi(t_n) + O((\sigma h)^{p+2}) = \sigma^{p+1}h^{p+1}\phi(t_n) + O(h^{p+2})$$
$$= \sigma^{p+1}est + O(h^{p+2}) \tag{2.11}$$

The largest step size that we predict will pass the error test corresponds to choosing $\sigma$ so that $|\sigma^{p+1}est| \approx \tau$. This step size is

$$h\left(\frac{\tau}{|est|}\right)^{1/(p+1)}$$

The solver is required to find a step size for which the magnitude of the estimated local error is no larger than the tolerance, so it must keep trying until it succeeds or gives up. It might give up because it has done too much work. It also might give up because it finds that it needs a step size too small for the precision of the computer. This is much like asking for an impossible relative accuracy, an issue discussed in Chapter 1. On the other hand, if the step is a success and the local error is rather smaller than necessary, we might increase the step size for the *next* step. This makes the computation more efficient because larger step sizes mean that we reach the end of the interval of integration in fewer steps. The same recipe can be used to estimate the step size that might be used on the next step: We predict that the error of a step of size $\sigma h$ taken from $t_{n+1}$ would be

$$u(t_{n+1} + \sigma h) - y_{n+2} = (\sigma h)^{p+1}\phi(t_{n+1}) + O(h^{p+2})$$
$$= \sigma^{p+1}h^{p+1}\phi(t_n) + O(h^{p+2})$$
$$= \sigma^{p+1}est + O(h^{p+2}) \tag{2.12}$$

In general terms, this is the way that popular codes select the step size, but we have omitted important practical details. For instance, how much the step size is increased or decreased must be limited because we cannot neglect the effects of higher-order terms (the "big oh" terms in equations (2.11) and (2.12)) when the change of step size is large. Also, several approximations are made in predicting the step size, so the estimate should not be taken too seriously. Because a failed step is relatively expensive and because the error estimate used to predict the step size may not be very reliable, the codes use a fraction of the predicted step size. Fractions like 0.8 and 0.9 are commonly used. The aim is to achieve the required accuracy without too great a chance of a failed step.

In the early days of numerical computing, when it was not understood how to estimate and control the local error, a constant step size was used. This approach is still seen today, but estimation and control of the local error is extremely important in practice – so important that it is used in all the computations of this book except when we wish to make a specific point about constant–step-size integration. Estimation and control of the local error is what gives us some confidence that we have computed a meaningful approximation to the solution of the IVP. Moreover, estimating the local error is not expensive. Generally it more than pays for itself because the step size is then not restricted to the smallest necessary to resolve the behavior of the solution over the whole interval of integration or to ensure stability of the integration. Indeed, it is impractical to solve many IVPs, including all stiff problems, with constant step size. Recall the proton transfer problem of Section 1.4. Its solution has a boundary layer of width about $10^{-10}$. Clearly we must use some steps of this general size to resolve the boundary layer, but the problem is posed on an interval of length about $10^6$. If we were to use a constant step size for the whole integration, we would need something like $10^{16}$ steps to solve this problem! This would take an impractically long time even on today's fastest computers, and even if it were possible to perform the calculation, the numerical solution would be dominated by the cumulative effects of roundoff error. In our solution of this problem for Figures 1.6 and 1.7, the integration required only $10^2$ steps. The step sizes ranged in size from $7 \cdot 10^{-14}$ in the boundary layer to $4 \cdot 10^4$ where the solution is slowly varying. The IVP was solved in a few seconds on a 433-MHz Pentium II PC and there were no obvious effects of roundoff error.

### Runge–Kutta Methods

We have now a brief yet nearly complete description of solving quadrature problems in the manner of a modern explicit Runge–Kutta code for IVPs. General IVPs are handled in much the same way, so mostly we point out differences. Now the local solution satisfies the IVP

$$u' = f(t, u), \quad u(t_n) = y_n$$

Again we integrate to obtain

$$u(t_n + h) = y_n + \int_{t_n}^{t_n+h} f(x, u(x)) \, dx$$

The crucial difference is that now the unknown local solution appears on both sides of the equation. If we approximate the integral with a quadrature formula, we have

$$\int_{t_n}^{t_n+h} f(x, u(x)) \, dx = h \sum_{j=1}^{s} A_j f(t_{n,j}, u(t_{n,j})) + O(h^{p+1}) \tag{2.13}$$

where again we abbreviate $t_n + \alpha_j h = t_{n,j}$. This doesn't seem like much help because we don't know the intermediate values $u(t_{n,j})$. Before discussing the two basic approaches to dealing with this problem, we consider some important examples of formulas for which there is no intermediate value.

The first (left) rectangle approximation to the integral is

$$\int_{t_n}^{t_n+h} f(x, u(x))\, dx = hf(t_n, u(t_n)) + O(h^2) = hf(t_n, y_n) + O(h^2)$$

The corresponding formula

$$y_{n+1} = y_n + hf(t_n, y_n) \tag{2.14}$$

is known as the *(forward) Euler method*. In the context of solving time-dependent partial differential equations it is known as the *fully explicit method*. The approximation

$$y_{n+1} = u(t_n + h) + O(h^2)$$

so it is a first-order explicit Runge–Kutta formula. The second (right) rectangle approximation leads to the *backward Euler method*

$$y_{n+1} = y_n + hf(t_{n+1}, y_{n+1}) \tag{2.15}$$

which is known for PDEs as the *fully implicit method*. Here we see a major difficulty that is not present for quadrature problems – the new approximate solution $y_{n+1}$ is defined implicitly as the solution of a set of algebraic equations. This formula is no more accurate than the forward Euler method, so why would we bother with the expense of evaluating an implicit formula? One answer is to overcome stiffness. As it happens, the backward Euler method is the lowest-order member of the family of backward differentiation formulas (BDFs) that we derive from a different point of view in the next section. The member of this family that is of order $k$ is denoted by BDF$k$, so the backward Euler method is also known as BDF1.

The trapezoidal rule

$$y_{n+1} = y_n + h\left[\tfrac{1}{2} f(t_n, y_n) + \tfrac{1}{2} f(t_{n+1}, y_{n+1})\right] \tag{2.16}$$

is a second-order implicit Runge–Kutta formula implemented in the MATLAB IVP solver `ode23t`. In the context of PDEs it is called the *Crank–Nicolson method*. Notice that the trapezoidal rule treats the solution values $y_n$ and $y_{n+1}$ in the same way. A formula like this is said to be *symmetric*. There is a direction of integration when solving IVPs, but usually not when solving BVPs. Because symmetric formulas do not have a preferred direction, they are widely used to solve BVPs.

Returning to the issue of intermediate values, suppose that we already have a formula that we can use to compute approximations $y_{n,j} = u(t_{n,j}) + O(h^p)$. Stating this assumption about the accuracy more formally, we assume there is a constant $C$ such that

$$\|u(t_{n,j}) - y_{n,j}\| \leq Ch^p$$

Along with our assumption that the ODE function $f$ satisfies a Lipschitz condition with constant $L$, this implies that

$$\|f(t_{n,j}, u(t_{n,j})) - f(t_{n,j}, y_{n,j})\| \leq L\|u(t_{n,j}) - y_{n,j}\| \leq LCh^p \qquad (2.17)$$

If we replace the function evaluations $f(t_{n,j}, u(t_{n,j}))$ with the computable approximations $f(t_{n,j}, y_{n,j})$, then a little manipulation of (2.13) and (2.17) shows that

$$\int_{t_n}^{t_n+h} f(x, u(x))\, dx = h \sum_{j=1}^{s} A_j f(t_{n,j}, y_{n,j}) + O(h^{p+1})$$

In this way we arrive at a formula

$$y_{n+1} = y_n + h \sum_{j=1}^{s} A_j f(t_{n,j}, y_{n,j})$$

We see that if we already know formulas that can be used to compute intermediate values $y_{n,j}$ accurate to $O(h^p)$, then we have constructed a formula to compute an approximate solution $y_{n+1}$ accurate to $O(h^{p+1})$.

There are two basic approaches to choosing the formulas for computing intermediate values. One is to use formulas of the same form as that for computing $y_{n+1}$. This leads to an implicit Runge–Kutta formula, a system of algebraic equations that generally involves computing simultaneously the approximate solution $y_{n+1}$ and the intermediate values $y_{n,j}$ for $j = 1, 2, \ldots, s$. To be useful for the solution of stiff IVPs, a formula must be implicit to some degree, but it is not necessary that all the $y_{n,j}$ be computed simultaneously. The MATLAB IVP solver `ode23tb` is based on an implicit formula that computes the intermediate values one at a time. If we choose explicit formulas for all the $y_{n,j}$, we obtain an explicit formula for $y_{n+1}$. Explicit formulas are popular for the solution of nonstiff problems. The MATLAB IVP solvers `ode23` and `ode45` are based on formulas of this kind.

## Explicit Runge–Kutta Formulas

Using the explicit forward Euler method, we can form the intermediate values needed for any quadrature formula with $p = 2$ to obtain a formula for which $u(t_n + h) = y_{n+1} + O(h^3)$. For example, if the quadrature formula is the trapezoidal rule, we obtain a second-order method called *Heun's method*:

$$
\begin{aligned}
y_{n,1} &= y_n + hf(t_n, y_n) \\
y_{n+1} &= y_n + h\left[\tfrac{1}{2}f(t_n, y_n) + \tfrac{1}{2}f(t_{n+1}, y_{n,1})\right]
\end{aligned}
\qquad (2.18)
$$

Exercise 2.10 asks you to solve an IVP with this formula. In Exercise 2.2 you are asked to construct a formula of the same order of accuracy using the midpoint rule instead of the trapezoidal rule. Using Heun's method, we can produce the intermediate values needed for any quadrature formula with order $p = 3$ to obtain a formula of order 4, and so forth. Because we can construct interpolatory quadrature formulas of any order, we see now how to construct an explicit Runge–Kutta formula of any order by this "bootstrapping" technique.

When we take account of the intermediate values, we find that the formulas resulting from the construction just outlined have the form of an explicit recipe that starts with

$$y_{n,1} = y_n, \qquad f_{n,1} = f(t_n, y_{n,1}) \tag{2.19}$$

and then, for $j = 2, 3, \ldots, s$, forms

$$y_{n,j} = y_n + h_n \sum_{k=1}^{j-1} \beta_{j,k} f_{n,k}, \qquad f_{n,j} = f(t_n + \alpha_j h_n, y_{n,j}) \tag{2.20}$$

and finishes with

$$y_{n+1} = y_n + h_n \sum_{j=1}^{s} \gamma_j f_{n,j} \tag{2.21}$$

The values (2.19) are the degenerate case $j = 1$ in (2.20) with $\alpha_1 = 0$, but they are written separately here to remind us that the method starts with the value $u(t_n) = y_n$ and slope

$$u'(t_n) = f(t_n, u(t_n)) = f(t_n, y_n)$$

of the local solution at the beginning of the step. A useful measure of the work involved in evaluating such an explicit formula is the number of evaluations of the function $f(x, y)$, that is, the number of *stages* $f_{n,j}$. Here the number of stages is $s$.

In principle we can work out the order of the formula given by equations (2.19), (2.20), and (2.21) just as we did with the formula (2.7) for quadrature problems. Indeed, if we apply the formula to a quadrature problem, then the conditions (2.8) may be stated in the present notation as

$$\frac{1}{k} = \sum_{j=1}^{s} \gamma_j \alpha_j^{k-1}, \quad k = 1, 2, \ldots, p \tag{2.22}$$

The conditions on the coefficients of a Runge–Kutta formula for it to be of order $p$ are called the *equations of condition*. The equations (2.22) are a subset and so they are necessary, but they are certainly not sufficient. Two matters complicate the argument in the general case. One is the presence of the solution $u(t)$ in the derivative $f(t, u(t))$, and the other is that $u(t)$ is a vector. To understand this better, let's look at the first nontrivial term in expanding $u(t_n + h)$ about $t_n$. In the Taylor series

$$u(t_n + h) = u(t_n) + u'(t_n)h + u''(t_n)\frac{h^2}{2} + \cdots$$

the IVP provides us immediately with $u(t_n) = y_n$ and $u'(t_n) = f(t_n, u(t_n)) = f(t_n, y_n)$. The next term is more complicated. If there are $d$ equations, component $i$ of the local solution satisfies

$$u'_i(t) = f_i(t, u_1(t), u_2(t), \ldots, u_d(t))$$

It is then straightforward to obtain

$$u''_i(t) = \frac{\partial f_i}{\partial t} + \sum_{j=1}^{d} \frac{\partial f_i}{\partial u_j} \frac{du_j}{dt} = \frac{\partial f_i}{\partial t} + \sum_{j=1}^{d} \frac{\partial f_i}{\partial u_j} f_j$$

In order to deal with formulas of even moderate orders, we clearly need to develop ways of making the manipulations easier. It would simplify the expressions considerably if the function $f(t, y)$ did not depend on $t$. Such a problem is said to be in *autonomous* form. Because it is always possible to obtain an equivalent IVP in autonomous form, the theory often assumes it (though most codes do not). We digress for a moment to show the usual way this is done for the IVP

$$\frac{dy}{dt} = f(t, y), \quad y(a) = A$$

on the interval $a \le t \le b$. If we change to the new independent variable $x = t$ and make $t$ a dependent variable, then

$$\frac{dY}{dx} = \frac{d}{dx}\begin{pmatrix} y \\ t \end{pmatrix} = \begin{pmatrix} f(t, y) \\ 1 \end{pmatrix} = F(Y), \quad Y(a) = \begin{pmatrix} A \\ a \end{pmatrix}$$

on $a \le x \le b$. This approach is convenient because all standard methods integrate the equation for $t$ exactly. Returning now to the derivation of methods, it is clear that a more powerful notation is needed as well as recursions for computing derivatives. This is all rather technical and we have no need for the details, so we take the special case of quadrature as representative. To gain an appreciation of the general case, Exercise 2.3 asks you to work out the details for a second-order explicit Runge–Kutta formula.

We have seen how to construct explicit Runge–Kutta formulas of any order, but generally the formulas constructed this way are not very efficient. Much effort has been devoted to finding formulas that yield a given order $p$ with as few stages as possible. For orders $p = 1, 2, 3$, and 4, the minimum number of stages is $s = p$; for order $p = 5$, it is $s = 6$. The minimum number of stages is interesting but not as important as it might seem. Extra stages can be used to find a more accurate formula. With a more accurate formula you can take larger steps, enough larger that you might be able to solve a problem with fewer

Table 2.1: *Butcher tableau for Runge–Kutta formulas.*

$$
\begin{array}{c|c}
\alpha & \beta \\
\hline
& \gamma
\end{array}
$$

Table 2.2: *The Euler–Heun (1, 2) pair.*

$$
\begin{array}{c|cc}
0 & & \\
1 & 1 & \\
\hline
& 1 & \\
\hline
& \frac{1}{2} & \frac{1}{2}
\end{array}
$$

overall evaluations of $f(t, y)$ even though each step is more expensive. Besides this basic point, the issue is not the cost of evaluating a formula by itself but rather the cost of evaluating a pair of formulas for taking a step and estimating the local error. The argument we made in deriving the estimate (2.9) of the local error was not restricted to quadrature problems. Deriving pairs of formulas that share many of their stages is a challenging task. We have remarked that at least $s = 6$ stages are required for a formula of order $p = 5$. Pairs of formulas of orders 4 and 5, denoted a $(4, 5)$ pair, are known that require a total of six stages. For example, a pair denoted F$(4, 5)$ due to Fehlberg (1970) using a total of six stages is in wide use. The popular $(4, 5)$ pair due to Dormand and Prince (1980), called DOPRI5, that is implemented in `ode45` has seven stages. The additional stage is used to improve the formula, and in tests DOPRI5 has proved somewhat superior to F$(4, 5)$.

The coefficients defining a Runge–Kutta formula are commonly presented as in Table 2.1, a notation due to Butcher. For an explicit Runge–Kutta method, all entries on and above the diagonal of the matrix $\beta$ are zero and it is conventional not to display them. When presenting pairs of formulas, the vectors of coefficients $\gamma$ are presented one above the other in the tableau. We have already seen a simple example of a pair with a minimal number of stages, namely the $(1, 2)$ pair consisting of the Euler and Heun formulas. Table 2.2 displays the tableau for this pair.

The Bogacki–Shampine (1989) BS$(2, 3)$ pair implemented in `ode23` is displayed in Table 2.3. The layout is a little different because the BS$(2, 3)$ pair exemplifies a technique

Table 2.3: *The* BS(2, 3) *pair.*

| | | | | |
|---|---|---|---|---|
| 0 | | | | |
| $\frac{1}{2}$ | $\frac{1}{2}$ | | | |
| $\frac{3}{4}$ | 0 | $\frac{3}{4}$ | | |
| 1 | $\frac{2}{9}$ | $\frac{1}{3}$ | $\frac{4}{9}$ | |
| | $\frac{7}{24}$ | $\frac{1}{4}$ | $\frac{1}{3}$ | $\frac{1}{8}$ |

called First Same As Last (FSAL). For an FSAL formula, the First stage of the next step is the Same As the Last stage of the current step. The last line of the table displays the coefficients $\gamma$ for the second-order formula of four stages. The line just above it contains the coefficients for the last stage, which by construction is a third-order formula of three stages. The way this works is that you form a third-order result $y_{n+1}$ with three stages, evaluate the fourth stage $f(t_{n+1}, y_{n+1})$, and then form the second-order result. This pair was designed to be used with local extrapolation, which is to say that the integration is to be advanced with the third-order result, $y_{n+1}$, as the approximate solution. The stage formed for the evaluation of the second-order formula and the error estimate is the first stage of the next step. In this way we get a stage for "free" if the step is accepted. In practice, most steps are accepted so this pair costs little more than a pair involving just three stages. Clearly, deriving pairs constrained to have the FSAL property is a challenge, but some popular pairs are of this form. One such is the seven-stage DOPRI5 pair mentioned earlier as the pair implemented in `ode45`. In practice it costs little more than the minimum of six stages per step that are required for any fifth-order formula. The widely used Fortran 77 package `RKSUITE` (Brankin, Gladwell, & Shampine 1993) and its close relative the Fortran 90 package `rksuite_90` (Brankin & Gladwell 1994) implement a (4, 5) FSAL pair that has still another stage. The topic of formula pairs is taken up in Exercise 2.4.

## Continuous Extensions

We have been discussing the approximation of $y(t)$ at points $t_0 < t_1 < \cdots$, but for some purposes it is valuable to have an approximation for all $t$. For example, plot packages draw straight lines between data points. Runge–Kutta formulas of moderate to high order take such long steps that it is quite common for straight line segments to be noticeable and distracting in plots of solution components. Figure 4.28 and 4.29 of Bender and Orszag (1999) provide good examples of this in the literature, and Exercise 2.21 provides another example. To plot a smooth graph, we need an inexpensive way of approximating the solution between mesh points. This is often called *dense output* in the context of Runge–Kutta methods.

Continuous extensions are a relatively recent development in the theory of Runge–Kutta methods. The idea is that, after taking a step from $t_n$ to $t_n + h$, the stages used in computing the solution and the error estimate (and perhaps a few additional stages) are used to determine a polynomial approximation to $u(t)$ throughout $[t_n, t_n + h]$. The scheme used in ode23 is particularly simple. At the beginning of the step we have approximations $y_n$ and $y_n' = f(t_n, y_n)$ to the value and slope of the solution there. At the end of the step we compute an approximation $y_{n+1}$. The BS(2, 3) pair also evaluates an approximation to the slope $y_{n+1}' = f(t_{n+1}, y_{n+1})$ because it is FSAL. (This information is readily available for any explicit Runge–Kutta formula because this slope is always the first stage of the next step.) With approximations to value and slope at both ends of an interval, we can use cubic Hermite interpolation to the solution and its slope at both ends of the current step. Using interpolation theory, it can be shown that this cubic interpolating polynomial approximates the local solution as accurately at all points of $[t_n, t_{n+1}]$ as $y_{n+1}$ approximates it at $t_{n+1}$. The cubic Hermite interpolant on $[t_{n-1}, t_n]$ has the same value and slope at $t_n$ as the interpolant on $[t_n, t_{n+1}]$, so this construction provides a piecewise-cubic polynomial $S(t) \in C^1[a, b]$. This approximation underlies the dde23 code for solving delay differential equations that we study in Chapter 4.

The interpolation approach to continuous extensions of Runge–Kutta formulas is valuable but somewhat limited. For each $\sigma \in [0, 1]$, an interpolant evaluated at $t_n + \sigma h$ can be viewed as a Runge–Kutta formula for taking a step of size $\sigma h$ from $t_n$. Another way to proceed is to derive such a family of formulas directly. The trick is to find a formula for taking a step of size $\sigma h$ that shares as many stages as possible with the formula that we use to step to $t_n + h$. The new formula depends on $\sigma$ but these stages do not, and any additional stages needed to achieve the desired order should also not depend on $\sigma$. It is possible to derive such families of formulas with coefficients that are polynomials in $\sigma$. The continuous extension of ode45 was derived in this way. No extra stages are needed to form its polynomial approximation of fourth order, but the piecewise-polynomial interpolant $S(t)$ is only continuous on the interval $[a, b]$. The DOPRI5 formula of ode45 allows the solver to take such large steps that by default it evaluates the continuous extension at four equally spaced points in the span of every step and returns them along with the approximations computed directly by the formula. Generally these additional solution values are sufficient to provide a smooth graph. An option is provided for increasing the number of output points if necessary.

■  **EXERCISE 2.1**

Using the equations of condition (2.8) for quadrature problems, verify that the following methods are of order 4.

- Simpson's method is based on Simpson's quadrature formula, also known as the three-point Lobatto formula,

$$\int_a^b f(x)\,dx \simeq \frac{b-a}{6}\left[f(a) + 4f\left(\frac{a+b}{2}\right) + f(b)\right]$$

- The two-point Gaussian quadrature method is based on the formula

$$\int_a^b f(x)\,dx \simeq \frac{b-a}{2}\left[f\left(\frac{a+b}{2} - \frac{b-a}{2\sqrt{3}}\right) + f\left(\frac{a+b}{2} + \frac{b-a}{2\sqrt{3}}\right)\right]$$

### ■ EXERCISE 2.2

Use Euler's method and the midpoint rule to derive a two-stage, second-order, explicit Runge–Kutta method.

### ■ EXERCISE 2.3

To understand better the equations of condition, derive the three equations for a formula of the form

$$y_{n+1} = y_n + h[\gamma_1 f_{n,1} + \gamma_2 f_{n,2}]$$

where

$$f_{n,1} = f(t_n, y_n)$$
$$f_{n,2} = f(t_n + \alpha_1 h, y_n + h\beta_{1,0} f_{n,1})$$

to be of second order. In the step from $(t_n, y_n)$, the result $y_{n+1}$ of the formula is to approximate the solution of

$$u' = f(t, u), \quad u(t_n) = y_n$$

at $t_{n+1} = t_n + h$. Expand $u(t_{n+1})$ and $y_{n+1}$ about $t_n$ in powers of $h$ and equate terms to compute the equations of condition. To simplify the expansions, do this for a scalar function $f(t, u)$.

### ■ EXERCISE 2.4

The explicit Runge–Kutta formulas

$$y_{n+1} = y_n + hf_{n,2}$$
$$y_{n+1}^* = y_n + \frac{h}{9}[2f_{n,1} + 3f_{n,2} + 4f_{n,3}]$$

of three stages

$$f_{n,1} = f(t_n, y_n)$$

$$f_{n,2} = f\left(t_n + \frac{h}{2}, y_n + \frac{h}{2}f_{n,1}\right)$$

$$f_{n,3} = f\left(t_n + \frac{3}{4}h, y_n + \frac{3}{4}hf_{n,2}\right)$$

are of order 2 and 3, respectively. State this $(2, 3)$ pair as a Butcher tableau. The local error of the lower-order formula is estimated by $est = y_{n+1}^* - y_{n+1}$. Suppose that the integration is to be advanced with the higher-order result (local extrapolation) and that you are given a relative error tolerance $\tau_r$ and absolute error tolerance $\tau_a$. This means that you will accept the step if

$$|est| \leq \tau_r |y_{n+1}^*| + \tau_a$$

What step size $h_{new} = \sigma h$ should you use if you must repeat the step because the estimated local error is too large? What step size should you use for the next step if the estimated local error is acceptable? Some solvers measure the error relative to $0.5(|y_n| + |y_{n+1}^*|)$ instead of $|y_{n+1}^*|$. Why might this be a good idea?

## 2.2.2 Methods with Memory

### Adams Methods

Once $t_n$ has been reached, we generally have available the previously computed solution values $y_n, y_{n-1}, \ldots$ and slopes $f_n = f(t_n, y_n)$, $f_{n-1} = f(t_{n-1}, y_{n-1})$, $\ldots$ that might be used in computing $y_{n+1}$. A natural way to exploit this information is a variation on the quadrature approach of the previous section. Recall that to approximate the local solution defined by

$$u' = f(t, u), \quad u(t_n) = y_n$$

we integrated to obtain

$$u(t_n + h) = y_n + \int_{t_n}^{t_n+h} f(x, u(x)) \, dx$$

Interpolating $s$ values $f_{n,j} = f(t_{n,j}, y_{n,j})$ with a polynomial $P(x)$ and then integrating, we derived a formula of the form

$$y_{n+1} = y_n + h \sum_{j=1}^{s} A_j f(t_{n,j}, y_{n,j})$$

We found that if the values $y_{n,j} \approx u(t_{n,j})$ are sufficiently accurate then this formula has order at least $s$. For one-step methods we required that the points $t_{n,j} \in [t_n, t_n + h]$. The question then was how to compute sufficiently accurate $y_{n,j}$. A natural alternative is to take $t_{n,j} = t_{n-j}$ because generally we already have sufficiently accurate approximations $y_{n,j} = y_{n-j}$ and, more usefully, $f_{n-j} = f(t_{n,j}, y_{n,j})$. This choice results in a family of explicit formulas called the *Adams–Bashforth formulas*. The lowest-order formula in this family is the forward Euler formula because it is the result of interpolating $f_n$ alone. In the present context it is called AB1. Interpolating $f_n$ and $f_{n-1}$ results after a little calculation in the second-order formula, AB2,

$$y_{n+1} = y_n + h_n\left[\left(1 + \frac{r}{2}\right)f_n - \left(\frac{r}{2}\right)f_{n-1}\right]$$

where $r = h_n/h_{n-1}$ (see Exercise 2.5). Note that, because we use previously computed values, the mesh spacing in the span of the memory appears explicitly in the coefficients. This is true in general, so it is necessary at each step to work out the coefficients of the formula. Techniques have been devised for doing this efficiently. For theoretical purposes, such formulas are often studied with the assumption that the step size is a constant $h$, in which case $r = 1$ and the formula for AB2 simplifies to

$$y_{n+1} = y_n + h\left[\tfrac{3}{2}f_n - \tfrac{1}{2}f_{n-1}\right]$$

The *Adams–Moulton formulas* arise in the same way except that the polynomial interpolates $f_{n+1}$. Because $f_{n+1} = f(t_{n+1}, y_{n+1})$ involves $y_{n+1}$, these formulas are implicit. The backward Euler method is the formula of order 1, and the trapezoidal rule is the formula of order 2. In this context they are called AM1 and AM2, respectively.

The accuracy of the Adams methods can be analyzed much as we did with Runge–Kutta methods. It turns out that the (implicit) Adams–Moulton formula of order $k$, AM$k$, is more accurate and more stable than the corresponding (explicit) Adams–Bashforth formula of order $k$, AB$k$. Which method is preferred depends on how much it costs to solve the nonlinear algebraic equation associated with the implicit formula. For nonstiff problems, implicit formulas are evaluated by what is called *simple iteration*.

We'll illustrate the iteration with the concrete example of AM1 because the general case is exactly the same and there are fewer terms to distract us. We solve iteratively the algebraic equations

$$y_{n+1} = y_n + hf(t_{n+1}, y_{n+1}) \tag{2.23}$$

for $y_{n+1}$. If the current iterate is $y_{n+1}^{[m]}$, the next is given by

$$y_{n+1}^{[m+1]} = y_n + hf(t_{n+1}, y_{n+1}^{[m]})$$

This operation is described as "correcting" the current iterate with an implicit formula that is called a *corrector formula*. Supposing that the algebraic equations (2.23) have a solution, it follows easily from the Lipschitz condition on $f$ that

$$\|y_{n+1} - y_{n+1}^{[m+1]}\| = \|hf(t_{n+1}, y_{n+1}) - hf(t_{n+1}, y_{n+1}^{[m]})\| \leq hL\|y_{n+1} - y_{n+1}^{[m]}\|$$

From this we see that, if $hL < 1$, the new iterate is closer to $y_{n+1}$ than the previous iterate and eventually we have convergence. This argument can be refined to prove that, for all sufficiently small step sizes $h$, the algebraic equations (2.23) have a solution and it is unique. (This is an example of a fixed-point argument that is common in applied mathematics.) The smaller the value of $h$, the faster this iteration converges. This is important because each iteration costs an evaluation of the ODE function $f$ and if many iterates are necessary then we might just as well use an explicit method with a smaller step size to achieve the same accuracy. On the other hand, for the sake of efficiency we want to use the largest step size that we can. An important way to reduce the number of iterations is to make a good initial guess $y_{n+1}^{[0]}$ for $y_{n+1}$. There is an easy and natural way to do this – predict the new value $y_{n+1}^{[0]}$ using an explicit formula, a *predictor formula*. For the implicit Adams–Moulton formula AM$k$, a natural predictor is an explicit Adams–Bashforth formula, either AB$k$ or AB$(k-1)$. Another important way to reduce the cost is to recognize that in practice it is not necessary to evaluate the implicit formula exactly, just well enough that the accuracy of the integration is not impaired. With considerable art in the implementation, a modern code like VODE (Brown, Byrne, & Hindmarsh 1989) that uses Adams–Moulton methods for nonstiff problems averages about two evaluations of $f$ per step. Simple iteration is practical for nonstiff problems because – in the classical situation, where $L(b-a)$ is not large – the requirement $Lh < 1$ cannot restrict the step size greatly. Choosing the step size to compute an accurate solution usually restricts the step size more than enough to ensure the rapid convergence of simple iteration.

There is an important variant of Adams methods that exemplifies a class of methods called *predictor–corrector methods*. A prediction is made with an Adams–Bashforth formula and then a *fixed* number of corrections is made with a corresponding Adams–Moulton formula. The most widely used methods of this kind correct only once. As it happens, Heun's method (2.18) is an example. The value $y_{n+1}^{[0]}$ is predicted with Euler's method, AB1. As we wrote Heun's method earlier, this value was called $y_{n,1}$. The rest of Heun's method is recognized as one correction with the trapezoidal rule (AM2) to give $y_{n+1}$, followed by evaluating $f(t_{n+1}, y_{n+1})$ for use on the next step. Proceeding in this way gives a PECE method (Predict–Evaluate $f$–Correct–Evaluate $f$). It would be natural to predict for AM2 with AB2, but this is not necessary and there are some advantages to using the lower-order predictor. It is not hard to show that predicting with a formula of order $k-1$ and correcting once with a formula of order $k$ results in a predictor–corrector formula of order $k$. The argument is much like the one used earlier for quadrature in general and Heun's method in particular. Although the predictor–corrector formula has the same order as the implicit formula used as corrector, the leading term in an expansion of the error is different. Exercise 2.7 provides an example of this. It is important to understand that a predictor–corrector method is an *explicit* method. Accordingly, it has qualitative properties that are in some respects quite different from the *implicit* formula that is used as the corrector; we'll mention one when we discuss stability. These different kinds of formulas are commonly confused because, for nonstiff problems, implicit formulas are evaluated by a prediction and correction process. The practical distinction is whether you use as many iterations as necessary to evaluate the implicit formula to a specified accuracy or use a fixed number of iterations. A difficulty with implicit methods is deciding reliably when the iteration to solve the formula has converged to sufficient accuracy, an issue not present with the explicit predictor–corrector methods. Although the two kinds of methods can differ substantially in certain situations, an Adams code that implements Adams methods as predictor–corrector pairs such as ODE/STEP, INTRP (Shampine & Gordon 1975) or ode113 performs much like an implementation of the Adams–Moulton implicit formulas such as in DIFSUB (Gear 1971) or VODE (Brown et al. 1989). The practice that you get implementing such formulas in Exercise 2.10 will help you understand better the distinction between predictor–corrector methods and implicit methods.

## BDF Methods

On reaching $t_n$, the *backward differentiation formula* of order $k$ (BDF$k$) approximates the solution $y(t)$ by the polynomial $P(t)$ that interpolates $y_{n+1}$ and the previously computed approximations $y_n, y_{n-1}, \ldots, y_{n+1-k}$. The polynomial is to satisfy the ODE at $t_{n+1}$, or (in a terminology that will be important for BVPs) it is to *collocate* the ODE at $t_{n+1}$. This requirement amounts to an algebraic equation for $y_{n+1}$,

$$P'(t_{n+1}) = f(t_{n+1}, P(t_{n+1})) = f(t_{n+1}, y_{n+1})$$

Now BDF1 results from linear interpolation at $t_{n+1}$ and $t_n$. The interpolating polynomial has a constant derivative, so the collocation equation is seen immediately to be

$$\frac{y_{n+1} - y_n}{h_n} = f(t_{n+1}, y_{n+1})$$

or, equivalently,

$$y_{n+1} - y_n = h_n f(t_{n+1}, y_{n+1})$$

Interpolating with a quadratic polynomial at the three points $t_{n+1}$, $t_n$, and $t_{n-1}$ leads in the same way to BDF2,

$$\left(\frac{1 + 2r}{1 + r}\right) y_{n+1} - (1 + r) y_n + \left(\frac{r^2}{1 + r}\right) y_{n-1} = h_n f(t_{n+1}, y_{n+1})$$

where $r = h_n/h_{n-1}$ (see Exercise 2.5). For a reason that we'll take up shortly, the BDFs are generally used with a constant step size $h$ for a number of steps. In this case, BDF2 is

$$\tfrac{3}{2} y_{n+1} - 2y_n + \tfrac{1}{2} y_{n-1} = hf(t_{n+1}, y_{n+1})$$

When the step size is a constant $h$, the Adams formulas and the BDFs are members of a class of formulas called *linear multistep methods* (LMMs). These formulas have the form

$$\sum_{i=0}^{k} \alpha_i y_{n+1-i} = h \sum_{i=0}^{k} \beta_i f(t_{n+1-i}, y_{n+1-i}) \tag{2.24}$$

In proving convergence for one-step methods, the propagation of the error made in a step could be bounded in terms of the stability of the IVP as the maximum step size tends to zero. Convergence is harder to prove when there is a memory because the error made in the current step depends much more strongly on the error made in preceding steps. This requires a shift of focus from approximating local solutions to approximating the global solution $y(t)$ and from the stability of the IVP to that of the numerical method. The *discretization error* or *local truncation error* ($lte_n$) of a linear multistep method is defined by

$$lte_n = \sum_{i=0}^{k} \alpha_i y(t_{n+1-i}) - h \sum_{i=0}^{k} \beta_i f(t_{n+1-i}, y(t_{n+1-i}))$$

$$= \sum_{i=0}^{k} \alpha_i y(t_{n+1-i}) - h \sum_{i=0}^{k} \beta_i y'(t_{n+1-i})$$

A straightforward expansion of the terms in Taylor series about $t_{n+1}$ shows that

$$lte_n = \sum_{j=0}^{\infty} C_j h^j y^{(j)}(t_{n+1}) \tag{2.25}$$

where

$$C_0 = \sum_{i=0}^{k} \alpha_i, \qquad C_1 = -\sum_{i=0}^{k} [i\alpha_i + \beta_i],$$

$$C_j = (-1)^j \sum_{i=1}^{k} \left[ \frac{i^j \alpha_i}{j!} + \frac{i^{j-1} \beta_i}{(j-1)!} \right], \quad j = 2, 3, \ldots$$

If an LMM is convergent, it is of order $p$ when its local truncation error is $O(h^{p+1})$. The expansion (2.25) shows that the formula is of order $p$ when $C_j = 0$ for $j = 0, 1, \ldots, p$. Furthermore,

$$lte_n = C_{p+1} h^{p+1} y^{(p+1)}(t_{n+1}) + \cdots = C_{p+1} h^{p+1} y^{(p+1)}(t_n) + \cdots \tag{2.26}$$

A couple of simple examples will be useful. Rearranging the Taylor series expansion

$$y(t_n) = y(t_{n+1}) - h y'(t_{n+1}) + \frac{h^2}{2} y''(t_{n+1}) + \cdots$$

UNIVERSITY OF HERTFORDSHIRE LRC

provides a direct proof that the local truncation error of the backward Euler method is

$$lte_n = -\frac{h^2}{2} y''(t_n + h) + \cdots = -\frac{h^2}{2} y''(t_n) + \cdots \tag{2.27}$$

We leave for Exercise 2.6 the computation that shows the local truncation error of the trapezoidal rule, AM2, to be

$$lte_n = -\frac{h^3}{12} y'''(t_n) + \cdots \tag{2.28}$$

A few details about proving convergence will prove illuminating. Suppose that we have an explicit LMM of the form

$$y_{n+1} = y_n + h \sum_{i=1}^{k} \beta_i f(t_{n+1-i}, y_{n+1-i})$$

a form that includes the Adams–Bashforth formulas. The solution of the ODE satisfies this equation with a small perturbation, the local truncation error,

$$y(t_{n+1}) = y(t_n) + h \sum_{i=1}^{k} \beta_i f(t_{n+1-i}, y(t_{n+1-i})) + lte_n$$

Subtracting the first equation from the second results in

$$y(t_{n+1}) - y_{n+1} = y(t_n) - y_n + h \sum_{i=1}^{k} \beta_i [f(t_{n+1-i}, y(t_{n+1-i})) - f(t_{n+1-i}, y_{n+1-i}] + lte_n$$

If the local truncation error is $O(h^{p+1})$, we can take norms and use the Lipschitz condition to obtain

$$\|y(t_{n+1}) - y_{n+1}\| \le \|y(t_n) - y_n\| + h \sum_{i=1}^{k} |\beta_i| L \|y(t_{n+1-i}) - y_{n+1-i}\| + Ch^{p+1}$$

This inequality involves errors at steps prior to $t_n$. The trick in dealing with them is to let

$$E_m = \max_{j \le m} \|y(t_j) - y_j\|$$

Using this quantity in the inequality, we have

$$\|y(t_{n+1}) - y_{n+1}\| \le (1 + h\mathcal{L}) E_n + Ch^{p+1}$$

where

$$L \sum_{i=0}^{k} |\beta_i| = \mathcal{L}$$

It then follows that

$$E_{n+1} \leq (1 + h\mathcal{L})E_n + Ch^{p+1}$$

This bound on the growth of the error in one step is just like the one we saw earlier for one-step methods. It is now easy to go on to prove that the error on the entire interval $[a, b]$ is $O(h^p)$. A small modification of the argument proves convergence for implicit methods that have the form of Adams–Moulton formulas. A similar argument can be used to prove convergence of predictor–corrector pairs like AB–AM in PECE form.

The error was bounded directly in the convergence proof just sketched, but often convergence is proven by first showing that the effects of small perturbations to the numerical solution are not amplified by more than a factor of $O(h^{-1})$. Because it takes $(b - a)h^{-1}$ steps to integrate from $a$ to $b$, we can think of this as stating roughly that the errors do no more than add up. A formula with this property is said to be *zero-stable*. At mesh points the solution of the ODE satisfies the formula with a small perturbation, the local truncation error, that is, $O(h^{p+1})$. Zero-stability then implies that the difference between the solution at mesh points and the numerical solution is $O(h^p)$ – that is, the method is convergent and of order $p$.

A convergence proof for general LMMs starts off in the same way, but now errors at previous steps appear without being multiplied by a factor of $h$. This makes the analysis more difficult, but more important is that, for some formulas, the errors at previous steps are amplified so much that the method does not converge. The classical theory of LMMs (see e.g. Henrici 1962, 1977) provides conditions on the coefficients of a formula that are necessary and sufficient for zero-stability. It turns out that zero-stable LMMs can achieve only about half the order of accuracy possible for the data used. One reason for giving our attention to the Adams formulas and BDFs is that they are zero-stable yet have about the highest order that is possible. The restriction on the order is necessary for a formula to be stable, but it is not sufficient. In fact, BDFs of orders 7 and higher are not zero-stable.

An example (Isaacson & Keller 1966, p. 381) of a linear multistep method that is not zero-stable is the explicit third-order formula

$$y_{n+1} + \frac{3}{2}y_n - 3y_{n-1} + \frac{1}{2}y_{n-2} - 3hf(t_n, y_n) = 0 \qquad (2.29)$$

In a numerical experiment we contrast it with AB3,

$$y_{n+1} - y_n - h\left[\frac{23}{12}f(t_n, y_n) - \frac{16}{12}f(t_{n-1}, y_{n-1}) + \frac{5}{12}f(t_{n-2}, y_{n-2})\right] = 0 \qquad (2.30)$$

These formulas use the same data and have the same order, so it is not immediately obvious that one is useful and the other is not. Nevertheless, the theory of LMMs tells us that,

Table 2.4: *Maximum error when* $h = 2^{-i}$.

| $i$ | AB3 | Formula (2.29) |
|---|---|---|
| 2 | 1.34e−003 | 9.68e−004 |
| 3 | 2.31e−004 | 6.16e−003 |
| 4 | 3.15e−005 | 1.27e+000 |
| 5 | 4.08e−006 | 6.43e+005 |
| 6 | 5.18e−007 | 2.27e+018 |
| 7 | 6.53e−008 | 4.23e+044 |
| 8 | 8.19e−009 | 2.27e+098 |
| 9 | 1.03e−009 | 1.03e+207 |
| 10 | 1.28e−010 | Inf |

as $h \to 0$, the zero-stable formula AB3 is convergent and formula (2.29) is not. The numerical experiment is to study how the maximum error depends on the step size $h$ when integrating the IVP $y' = -y$, $y(0) = 1$ over $[0, 1]$. Starting values are taken from the analytical solution. Table 2.4 displays the maximum error for step sizes $h = 2^{-i}$, $i = 2, 3, \ldots, 10$. It appears that AB3 is converging as $h \to 0$. Indeed, when $h$ is halved, the maximum error is divided by about 8, as we might expect of a third-order formula. It appears that formula (2.29) is not converging. The error is not so bad for the larger values of $h$ because only a few steps are taken after starting with exact values, but not many steps are needed for the error to be amplified to the extent that the approximate solutions are unacceptable. Exercise 2.8 asks you first to verify that formula (2.29) is of order 3 and then to perform this experiment yourself.

When the step size is varied, we must expect some restrictions on how fast it can change if we are to have stability and convergence as a maximum step size tends to zero. If the ratio of successive step sizes is uniformly bounded above, then stability and convergence can be proved for Adams methods much as we did for constant step size (Shampine 1994). In practice this condition is satisfied because the solvers limit the rate of increase of step size for reasons explained in Section 2.2.1. The BDFs are another matter. The classical theory for LMMs shows that the BDFs of order less than 7 are stable and convergent as a constant step size tends to zero. They are often implemented so that the solver uses one constant step size until it appears advantageous to change to a different constant step size. There are theoretical results that show stability and convergence of such an implementation, provided the changes of step size are limited in size and frequency. Unfortunately, the situation remains murky because the theoretical results do not account for changes as large and frequent as those seen in practice.

The numerical methods used in practice are all stable as a maximum step size tends to zero. But in a specific integration, are the step sizes small enough that the integration is stable? It is easy enough to write down expressions for the propagation of small perturbations to a numerical solution, but the expressions are so complicated that it is difficult to gain any insight. For guidance we turn to a standard analysis of the stability of the ODE itself. The idea is to approximate an equation in autonomous form, $y' = f(y)$, near a point $(t^*, y^*)$ by a linear equation with constant coefficients,

$$u' = f(y^*) + \frac{\partial f}{\partial y}(y^*)(u - y^*) \tag{2.31}$$

The approximating ODE is unstable if there are solutions starting near $(t^*, y^*)$ that spread apart rapidly. The difference of any two solutions of this linear equation is a solution of the homogeneous equation

$$v' = \frac{\partial f}{\partial y}(y^*)v$$

For simplicity it is usual to assume that the local Jacobian $\frac{\partial f}{\partial y}(y^*)$ is diagonalizable, meaning that there is a matrix $T$ of eigenvectors such that $T^{-1}\frac{\partial f}{\partial y}(y^*)T = \text{diag}\{\lambda_1, \lambda_2, \ldots, \lambda_d\}$. If we change variables to $w = T^{-1}v$, we find that the equations uncouple: component $j$ of $w$ satisfies $w'_j = \lambda_j w_j$. Each of these equations is an example of the (scalar) *test equation*

$$w' = \lambda w \tag{2.32}$$

with $\lambda$ an eigenvalue of the local Jacobian. Solving these equations shows that, if $\text{Re}(\lambda_j) > 0$ for some $j$, then $w_j(t)$ grows exponentially fast and so does $v(t)$. Because the difference between two solutions of (2.31) grows exponentially fast, the ODE is unstable. On the other hand, if $\text{Re}(\lambda_j) \leq 0$ for all $j$, then $w(t)$ is bounded and so is $v(t)$. The approximating equation (2.31) is then stable near $(t^*, y^*)$. The standard numerical methods can be analyzed in a similar way to find that the stability of the method depends on the numerical solution of the test equation. As we shall see, it is not hard to work out the behavior of the numerical solution of this equation when the step size is a constant $h$. With it we can answer the fundamental question: If $\text{Re}(\lambda_j) \leq 0$ for all $j$ so that the approximating ODE is stable near $(t^*, y^*)$, how small must $h$ be in order for the numerical solution to be stable? The theory of *absolute stability* that we have outlined involves a good many approximations, but it has proven to be quite helpful in understanding practical computation. We can also think of it as providing a necessary condition, for if a method does not do a good job of solving the test equation then it can be of only limited value for general equations.

Euler's method provides a simple example of the theory of absolute stability. Suppose $\text{Re}(\lambda) \leq 0$, so that all solutions of the test equation are bounded and the equation is stable. If we apply the forward Euler method to this equation, we find that

$$y_{n+1} = y_n + h\lambda y_n = (1 + h\lambda)y_n$$

Clearly it is necessary that $|1 + h\lambda| \leq 1$ for a bounded numerical solution. The set

$$S = \{|1 + z| \le 1, \operatorname{Re}(z) \le 0\}$$

is called the *(absolute) stability region* of the forward Euler method. If $h\lambda \in S$ then the numerical method is stable, just like the differential equation. If $h\lambda$ is not in $S$, the integration blows up. This is applied to more general problems by requiring that $h\lambda_j \in S$ for all the eigenvalues $\lambda_j$ of the local Jacobian $\frac{\partial f}{\partial y}$. Again we caution that a good many approximations are made in the theory of absolute stability. Nevertheless, experience tells us that it provides valuable insight.

Whatever the value of $\lambda$, the forward Euler method is stable for the test equation for all sufficiently small $h$, as we already knew because the method converges. Notice that $L = |\lambda|$ is a Lipschitz constant for the test equation. In the classical situation where $L(b - a)$ is not large, any restriction on the step size necessary to keep the computation stable cannot be severe. But what if $\operatorname{Re}(\lambda) < 0$ and $|\lambda|(b - a) \gg 1$? In this case the differential equation is stable – indeed, extremely so because solutions approach one another exponentially fast – but stability of the forward Euler method requires that $h$ be not much larger than $|\lambda|^{-1}$.

The step size that might be used in practice is mainly determined by two criteria, accuracy and stability. Generally the step size is determined by the accuracy of a formula, but for *stiff* problems, it may be determined by stability. To understand better what stiffness is, let us consider the solution of the IVP

$$y' = -100y + 10, \quad y(0) = 1 \tag{2.33}$$

on (say) $0 \le t \le 10$. Exercise 2.11 asks you to determine the largest step size for which the leading term of the local truncation error of the forward Euler method is smaller in magnitude than a specified absolute error, but the qualitative behavior of the step size is clear. The analytical solution

$$y(t) = \frac{1}{10} + \frac{9}{10}e^{-100t}$$

shows that there is an initial period of very rapid change, called a *boundary layer* or *initial transient*. In this region the method must use a small step size to approximate the solution accurately, a step size so small that the integration is automatically stable. The problem is not stiff for this method in the initial transient. After a short time $t$, the solution $y(t)$ is very nearly constant, so an accurate solution can be obtained with a large step size. However, the step size must satisfy $|1 + h(-100)| \le 1$ if the computation is to be stable. This is frustrating. The solution is easy to approximate, but if we try a step size larger than $h = 0.02$, the integration will blow up. Modern codes that select the step size automatically do not blow up: if the step size is small enough that the integration is stable, the code increases the step size because the solution is easy to approximate. When the step size is increased to the point that it is too large for stability, errors begin to grow. When the error

Table 2.5: *Solution of a mildly stiff IVP.*

| Solver | AE | ME | SS | FS |
|--------|------|------|-----|-----|
| ode45 | 1.0e−1 | 1.1e−1 | 303 | 26 |
|  | 1.0e−2 | 1.1e−2 | 304 | 26 |
|  | 1.0e−3 | 1.1e−3 | 307 | 19 |
|  | 1.0e−4 | 1.1e−4 | 309 | 19 |
| ode15s | 1.0e−1 | 3.4e−2 | 23 | 0 |
|  | 1.0e−2 | 8.2e−3 | 29 | 0 |
|  | 1.0e−3 | 1.1e−3 | 39 | 0 |
|  | 1.0e−4 | 1.6e−4 | 65 | 0 |

*Key:* AE, absolute error tolerance; ME, maximum error; SS, number of successful steps; FS, number of failed steps.

becomes too large, the step is a failure and the step size is reduced. This is repeated until the step size is small enough that the integration is again stable and the cycle repeats. The code computes an accurate solution, but the computation is expensive – not only because a small step size is necessary for stability but also because there are many failed steps. Exercise 2.15 provides an example.

A numerical experiment with the IVP (2.33) is instructive. Because $0 < y(t) \leq 1$, a relative error tolerance of $10^{-12}$ and a modest absolute error tolerance is effectively a pure absolute error control. Table 2.5 shows what happens when we solve this IVP with the explicit Runge–Kutta code ode45 for a range of absolute error tolerances. This code is intended for nonstiff problems, but it can solve the IVP because its error control keeps the integration stable. Further, the cost of solving the IVP is tolerable because it is only mildly stiff. Notice that there is a relatively large number of failed steps. Also, the number of successful steps is roughly constant as the tolerance is decreased. This is characteristic of stiff problems because, for much of the integration, the step size is not determined by accuracy. The BDF code ode15s that is intended for stiff problems behaves quite differently. The number of successful steps depends on the tolerance because the step size is determined by accuracy. For this problem, ode15s does not have the step failures that ode45 does because of its finite stability region.

Stiff problems are very important in practice, so we must look for methods more stable than the forward Euler method. We don't need to look far. If we solve the test equation with the backward Euler method, the formula is

$$y_{n+1} = y_n + h\lambda y_{n+1}$$

and hence

$$y_{n+1} = \frac{1}{1 - h\lambda} y_n \qquad (2.34)$$

The stability region of the backward Euler method is then the set

$$S = \left\{ \left| \frac{1}{1 - z} \right| \leq 1, \ \mathrm{Re}(z) \leq 0 \right\}$$

which is found to be the whole left half of the complex plane. This is a property called *A-stability*. There appears to be no restriction on the step size for a stable integration with the backward Euler method. Many approximations were made in the stability analysis, so you shouldn't take this conclusion too seriously. That said, the backward Euler method does have excellent stability. This method is BDF1 and, similarly, the BDFs of orders 2 through 6 have stability regions that extend to infinity. Only the formulas of orders 1 and 2 are *A*-stable. The others are stable in a sector

$$\{z = re^{i\theta} \mid r > 0, \ \pi - \alpha < \theta < \pi + \alpha\}$$

that contains the entire negative real axis. This restricted version of *A*-stability is called *A($\alpha$)-stability*. As the order increases, the angle $\alpha$ becomes smaller and BDF7 is not stable for any angle $\alpha$; that is, it is not stable at all.

Like BDF1, the trapezoidal rule is *A*-stable. The predictor–corrector pair that consists of the forward Euler predictor and one correction with the trapezoidal rule is an explicit Runge–Kutta formula, Heun's method. It is not hard to work out the stability region of Heun's method and so learn that it is finite. In fact, *all* explicit Runge–Kutta methods have finite stability regions. This shows that an implicit method and a predictor–corrector pair using the same implicit method as a corrector can have important qualitative differences.

Exercise 2.9 takes up stability regions for some one-step methods and asks you to prove some of the facts just stated. Exercise 2.16 takes up the computation of stability regions for LMMs. For stability all we ask is that perturbations of the numerical solution not grow. However, if $\mathrm{Re}(\lambda) < 0$ then perturbations of the ODE itself decay exponentially fast. It would be nice if the numerical method had a similar behavior. The expression (2.34) for BDF1 shows that perturbations are damped strongly when $h\,\mathrm{Re}(\lambda) \ll -1$. An *A($\alpha$)*-stable formula with this desirable property is said to be *L($\alpha$)-stable*. An attractive feature is that the convergent BDFs are *L($\alpha$)*-stable. For the trapezoidal rule, perturbations are barely damped when $h\,\mathrm{Re}(\lambda) \ll -1$; the formula is *A*-stable, but not *L($\alpha$)*-stable.

Interestingly, the BDFs are stable for *all* sufficiently large $|h\lambda|$. If $\mathrm{Re}(\lambda) > 0$ and a BDF is stable for $h\lambda$, then the solution of the test equation grows and the numerical solution decays. Of course the numerical solution has the right qualitative behavior for all sufficiently small $h\lambda$ because the method is convergent, but if $\mathrm{Re}(\lambda) > 0$ and $h$ is too large then you may not like the heavy damping of this formula. Exercises 2.12 and 2.13 take up implications of this. Exercise 2.14 considers how the implementation affects stability.

What's the catch? Why have we even been talking about methods for nonstiff problems? Well, all the methods with infinite stability regions are implicit. Earlier, when we derived the backward Euler method as AM1, we talked about evaluating it by simple iteration. Unfortunately, the restriction on the step size for convergence of simple iteration is every bit as severe as that due to stability for an explicit method. To see why this is so, let us evaluate the backward Euler method by simple iteration when solving the test equation. The iteration is

$$y_{n+1}^{[m+1]} = y_n + h\lambda y_{n+1}^{[m]}$$

and

$$|y_{n+1} - y_{n+1}^{[m+1]}| = |h\lambda||y_{n+1} - y_{n+1}^{[m]}|$$

Clearly we must have $|h\lambda| < 1$ for convergence. This is not important when solving nonstiff problems, but it is not acceptable when solving stiff problems. For the example (2.33) of a stiff IVP, $h$ must be less than 0.01 for convergence. This restriction of the step size due to simple iteration when using the backward Euler method is worse than the one due to stability when using the forward Euler method! To solve stiff problems we must resort to a more powerful way of solving the algebraic equations of an implicit method. In the case of BDFs, these equations have the form

$$y_{n+1} = h\gamma f(t_{n+1}, y_{n+1}) + \psi \tag{2.35}$$

Here $\gamma$ is a constant that is characteristic of the method and $\psi$ lumps together terms involving the memory $y_n, y_{n-1}, \ldots$. As with simple iteration for nonstiff problems, it is important to make a good initial guess $y_{n+1}^{[0]}$. This can be achieved by interpolating $y_n, y_{n-1}, \ldots$ with a polynomial $Q(t)$ and taking $y_{n+1}^{[0]} = Q(t_{n+1})$. A simplified Newton (chord) method can then be used to solve the algebraic equations iteratively. The equations are linearized approximately as

$$y_{n+1}^{[m+1]} = \psi + h\gamma[f(t_{n+1}, y_{n+1}^{[m]}) + J(y_{n+1}^{[m+1]} - y_{n+1}^{[m]})]$$

Here

$$J \approx \frac{\partial f}{\partial y}(t_{n+1}, y_{n+1})$$

This way of writing the iteration shows clearly the approximate linearization, but it is very important to organize the computation properly. The next iterate should be computed as a correction $\Delta_m$ to the current iterate. A little manipulation shows that

$$(I - h\gamma J)\Delta_m = \psi + h\gamma f(t_{n+1}, y_{n+1}^{[m]}) - y_{n+1}^{[m]} \tag{2.36}$$

$$y_{n+1}^{[m+1]} = y_{n+1}^{[m]} + \Delta_m$$

The iteration matrix $I - h\gamma J$ is very ill-conditioned when the IVP is stiff. Exercise 2.20 explains how to modify `ode15s` so that you can monitor the condition of the iteration matrix in the course of an integration and then asks you to verify this assertion for one IVP. When solving a very ill-conditioned linear system, it may be that only a few of the leading digits of the solution are computed correctly. Accordingly, if we try to compute the iterates $y_{n+1}^{[m+1]}$ directly, we may not be able to compute more than a few digits correctly. However, if we compute a few correct digits in each increment $\Delta_m$, then more and more digits will be correct in the iterates $y_{n+1}^{[m+1]}$. Notice that the right-hand side of the linear system (2.36) is the residual of the current iterate. It is a measure of how well the current iterate satisfies the algebraic equations (2.35). By holding fixed the approximation $J$ to the local Jacobian, an $LU$ factorization (that is, an $LU$ decomposition) of the iteration matrix $I - h_n\gamma J$ can be computed and then used to solve efficiently all the linear systems of the iteration at the point $t_{n+1}$. Forming and factoring $I - h_n\gamma J$ is relatively expensive, so BDF codes typically hold the step size constant and use a single $LU$ factorization for several steps. Only when the iterates do not converge sufficiently fast or when it is predicted that a considerably larger step size might be used (perhaps with a change of order, i.e., with a different formula) is a new iteration matrix formed and factored.

Evaluating an implicit formula by solving the equation with a simplified Newton iteration can be expensive. In the first place, we must form and store an approximation $J$ to the local Jacobian. We then must solve a linear system with matrix $I - h\gamma J$ for each iterate. It can be expensive to approximate Jacobians, so the codes try to minimize the number of times this is done. A new iteration matrix must be formed when the step size changes significantly. The early codes for stiff IVPs formed a new approximation to the Jacobian at this time and overwrote it with the new iteration matrix because storage was at a premium. This is still appropriate when solving extremely large systems, but nowadays some solvers for stiff IVPs, including those of MATLAB, save $J$ and reuse it in the iteration matrix for as long as the iteration converges at an acceptable rate. This implies that if such a solver is applied to a problem that is not stiff, very few Jacobians are formed. Along with the fast linear algebra of the MATLAB PSE, this makes the stiff solvers of MATLAB reasonably efficient for nonstiff problems of modest size.

Because it is so convenient, all the codes for stiff IVPs have an option for approximating Jacobians by finite differences. This is the default option for the MATLAB solvers. Suppose that we want to approximate the matrix $\frac{\partial f}{\partial y}(t^*, y^*)$. If $e^{(j)}$ is column $j$ of the identity matrix, then the vector $y^* + \delta_j e^{(j)}$ represents a change of $\delta_j$ in component $j$ of $y^*$. From the Taylor series

$$f(t^*, y^* + \delta e^{(j)}) = f(t^*, y^*) + \frac{\partial f}{\partial y}(t^*, y^*)\delta_j e^{(j)} + O(\delta_j^2)$$

we obtain an approximation to column $j$ of the Jacobian:

$$\frac{\partial f}{\partial y}(t^*, y^*)e^{(j)} \approx \delta_j^{-1}[f(t^*, y^* + \delta e^{(j)}) - f(t^*, y^*)]$$

For a system of $d$ equations, we can obtain an approximation to the Jacobian in this way by making $d$ evaluations of $f$. This is the standard method of approximating Jacobians, but the algorithms differ considerably in detail because it is hard to choose a good value for $\delta_j$. It must be small enough that the finite differences provide a good approximation to the partial derivatives but not so small that the approximation consist only of roundoff error. When the components of $f$ differ greatly in size, it may not even be possible to find one value of $\delta_j$ that is good for all components of the vector of partial derivatives. Fortunately, we do not need an accurate Jacobian, just an approximation that is good enough to achive acceptable convergence in the simplified Newton iteration. A few algorithms, including the numjac function of MATLAB, monitor the differences in the function values. They adjust the sizes of the increments $\delta_j$ based on experience in approximating a previous Jacobian and repeat the approximation of a column with a different increment when this appears to be necessary for an acceptable approximation. Numerical Jacobians are convenient and generally satisfactory, but the solvers are more robust and perhaps faster if you provide an analytical Jacobian.

For a large system of ODEs, it is typical that only a few components of $y$ appear in each equation. If component $j$ of $y$ does not appear in component $i$ of $f(t, y)$, then the partial derivative $\frac{\partial f_i}{\partial y_j}$ is zero. If most of the entries of a matrix are zero, the matrix is said to be *sparse*. By storing only the nonzero entries of a sparse Jacobian, storage is reduced from the square of the number of equations $d$ to a modest multiple of $d$. If the Jacobian is sparse then so is the iteration matrix. As with storage, the cost of solving linear systems by elimination can be reduced dramatically by paying attention to zero entries in the matrix. A clever algorithm of Curtis, Powell, & Reid (1974) provides a similar reduction in the cost of approximating a sparse Jacobian. By taking into account the known value of zero for most of the entries in the Jacobian, it is typically possible to approximate all the nonzero entries of several columns at a time. An important special case of a sparse Jacobian is one that has all its nonzero entries located in a band of diagonals. For example, if for all $i$ the entry $J_{i,j} = 0$ for all $j$ except possibly $j = i - 1$, $i$, and $i + 1$, we say that the matrix $J$ is *tridiagonal* and has a band width of 3. If there are $m$ diagonals in the band, only $m$ additional evaluations of $f$ are needed to approximate the Jacobian *no matter how many equations there are.* It is comparatively easy to obtain the advantages of sparsity when the matrix is banded, so all the popular solvers provide for banded Jacobians. Some, including all the solvers of MATLAB, provide for general sparse Jacobians. These algorithmic developments are crucial to the solution of large systems. Section 2.3.3 provides more information about this as well as some examples.

Each step in an implementation of BDFs that uses a simplified Newton iteration to evaluate the formulas is much more expensive – in terms of computing time and storage – than taking a step with a method intended for nonstiff problems using either simple iteration or no iteration at all. Accordingly, BDFs evaluated with a simplified Newton iteration are advantageous only when the step size would otherwise be greatly restricted by stability, that is, only when the IVP is stiff. For nonstiff problems the BDFs could be evaluated with simple iteration, but this is rarely done because the Adams–Moulton methods have a similar structure and are considerably more accurate.

It is unfortunate that stiff problems are not easily recognized. We do have a clear distinction between methods and implementations for stiff and nonstiff problems. A practical definition of a stiff problem is that it is a problem such that solving it with a method intended for stiff problems is much more efficient than solving it with a method intended for nonstiff problems. Exercises 2.15, 2.18, 2.34, and 2.36 will give you some experience with this. Insight may provide guidance as to whether a problem is stiff. The ODEs of a stiff problem must allow solutions that change on a scale that is small compared to the length of the interval of integration. Some physical problems are naturally expressed in terms of time constants, in which case a stiff problem must have some time constant that is small compared to the time interval of interest. The solution of interest must itself be slowly varying; Exercise 2.36 makes this point. Most problems that are described as stiff have regions where the solution of interest changes rapidly, such as a boundary layer or an initial transient. In such intervals the problem is not stiff because the solver must use a small step size to represent the solution accurately. The problem is stiff where the solution is easy to approximate and the requirement for stability dominates the choice of step size. Variation of the step size – so as to use step sizes appropriate both for the initial transient and for the smooth region – is clearly fundamental to the solution of stiff IVPs. The proton transfer and Robertson examples of Section 1.4 illustrate this.

## Error Estimation and Change of Order

Estimation of the local truncation error of linear multistep methods is generally considered to be easy compared to estimation of the local error for Runge–Kutta methods. It *is* easy to derive estimates – what's hard is to prove that they work. We begin by discussing how the local truncation error might be estimated. With the Adams methods, it is easy to take a step with formulas of two different orders. Indeed, we noted earlier that we could predict with $AB(k-1)$ and obtain a formula of order $k$ with a single correction using $AMk$. Just as with one-step methods, the error in the formula $AB(k-1)$ of order $k-1$ can be estimated by comparing the result of this formula to the result of order $k$ from the pair. When the size of this error is controlled and the integration is advanced with the more accurate and stable result of the predictor–corrector pair, we are doing local extrapolation. This approach is used in `ode113`. Another approach is based on the expansion (2.26) for the local truncation error when the step size is constant. It shows that the leading terms of the local truncation errors of the Adams–Bashforth and Adams–Moulton methods of the same order differ by a constant multiple. Using this fact, the step is taken with both AB$k$ and AM$k$ and the local truncation error is estimated by an appropriate multiple of the difference of the two results. The more accurate and stable result of AM$k$ is used to advance the integration. The error of this formula is being controlled, so local extrapolation is not done. In this approach the implicit Adams–Moulton methods are evaluated by simple iteration to a specified accuracy. The explicit Adams–Bashforth formula AB$k$ is used to start the iteration for computing AM$k$ and to estimate the local truncation error of AM$k$. To leading order, the local truncation errors of AM$k$ and the AB$k$–AM$k$ pair with one iteration are the same, so the same approach to local truncation error estimation can also be used for predictor–corrector implementations.

An approach that is particularly natural for BDFs is to approximate directly the leading term of the local truncation error. Recall that for BDF1 this is

$$lte_n = -\frac{h^2}{2} y''(t_n) + \cdots$$

The formulas are based on numerical differentiation, so it is natural to interpolate $y_{n+1}$, $y_n$, and $y_{n-1}$ with a quadratic polynomial $Q(t)$ and then use

$$est = -\frac{h^2}{2} Q''(t_n) \approx -\frac{h^2}{2} y''(t_n)$$

This is how the local truncation error is estimated for the BDFs in `ode15s`. Similarly, a cubic interpolant is used in `ode23t` to approximate the derivative in the local truncation error (2.28) of the trapezoidal rule. The backward Euler method and the trapezoidal rule are one-step methods, but previously computed approximate solutions are used for estimating local truncation errors. Previously computed solutions are also used to predict the solution of the algebraic equations that must be solved to evaluate these implicit methods. We see from this that, in practice, the distinction between implicit one-step methods and methods with memory may be blurred.

An interesting and important aspect of these estimates of the local truncation error is that it is possible to estimate the error that would have been made if a different order had been used. The examples of AM1 and AM2 show the possibility. Considering how the local truncation error is estimated, it is clear that – when taking a step with the second-order formula AM2 – we could estimate the local truncation error of the first-order formula AM1. At least mechanically, it is also clear that we could use more memorized values to estimate the error that would have been made with the higher-order formula AM3. This opens up the possibility of adapting the order (formula) to the solution so as to use larger step sizes. Modern Adams and BDF codes like `ode113` and `ode15s`, respectively, do exactly that. Indeed, the names of these functions indicate which members of the family are used: the Adams code `ode113` uses orders ranging from 1 to 13, and the BDF code

`ode15s` uses orders ranging from 1 to 5. (The final `s` of `ode15s` indicates that it is intended for stiff IVPs.) Both Adams–Moulton formulas and BDFs are implemented in the single variable–order code `DIFSUB` (Gear 1971) and also in its descendants such as `VODE` (Brown et al. 1989). Variation of the order plays another role in Adams and BDF codes. The lowest-order formulas are one-step, so they can be used to take the first step of the integration. Each step of the integration provides another approximate solution value for the memory, so the solver can consider raising the order and increasing the step size for the next step. There are many practical details, but the codes all increase the order rapidly until a value appropriate to the solution has been found. Starting the integration in this way is both convenient and efficient in a variable–step size, variable-order (VSVO) implementation, so all the popular Adams and BDF codes do it this way.

Deriving estimates of the local truncation error is easy for methods with memory, but justifying them is hard. The snag is this: When we derived the estimators, we assumed implicitly that the previously computed values and derivatives are exactly equal to the solution values $y(t_n)$, $y(t_{n-1})$, ... and their derivatives $y'(t_n)$, $y'(t_{n-1})$, ..., respectively; in practice, however, these previously computed values are in error, and these errors may be just as large *or even larger* than the error we want to estimate. The difficulty is especially clear when we contemplate estimating the local truncation error of a formula with order higher than the one(s) used to compute the memorized values. Justifying an estimator in realistic circumstances is difficult because it is not even possible without some regularity in the behavior of the error. So far we have discussed only the order of the error. Classic results in the theory of linear multistep methods show that, with certain assumptions, the error behaves in a regular way. This regularity can be used to justify error estimation, as was done first by Henrici (1962) and then more generally by Stetter (1973). Their results do not apply directly to modern variable-order Adams and BDF codes because they assume that the same formula is used all the time and that, if the step sizes are varied at all, they are varied in a regular way that is given a priori. Little is known about the behavior of the error when the order (formula) is varied in the course of the integration. The theory for constant order provides insight and there are a few results (Shampine 2002) that apply directly, but our theoretical understanding of modern Adams and BDF codes that vary their order leaves much to be desired.

## Continuous Extensions

Both Adams methods and BDFs are based on polynomial interpolants, so it is natural to use these interpolants as continuous extensions for the methods. An application of continuous extensions that is special to methods with memory is changing the step size. We have seen that there are practical reasons for working with a constant step size when solving stiff IVPs. An easy way to change to a different constant step size $H$ at $t_n$ is to approximate the solution at $t_n, t_n - H, t_n - 2H, \ldots$ by evaluating the interpolant at these points.

With these solution values at a constant mesh spacing of $H$, we can use a constant-step formula with step size $H$ from $t_n$ on. This is a standard technique for changing the step size when integrating with BDFs, and some codes use it also for Adams methods; DIFSUB is an example for both kinds of methods.

### ■ EXERCISE 2.5

To understand better the origin of Adams formulas and BDFs:

- Work out the details for deriving AB2; and
- Work out the details for deriving BDF2.

### ■ EXERCISE 2.6

Show that the local truncation error of AM2, the trapezoidal rule, is

$$lte_n = -\frac{h^3}{12}y^{(3)}(t_n) + \cdots$$

### ■ EXERCISE 2.7

Show that, when solving $y' = -y$, the error made in a step of size $h$ from $(t_n, y(t_n))$ with AM2 (the trapezoidal rule) is

$$y(t_n + h) - y_{n+1} = \frac{1}{12}h^3 y(t_n) + \cdots$$

(This is the local truncation error of AM2 for this ODE.) Show that the corresponding error for AB1–AM2 in PECE form (Heun's method) is

$$y(t_n + h) - y_{n+1} = -\frac{1}{6}h^3 y(t_n) + \cdots$$

Evidently the accuracy of a predictor–corrector method can be different from that of the corrector evaluated as an implicit formula.

### ■ EXERCISE 2.8

Verify that the formula (2.29) is of order 3. Do the numerical experiment described in the text that resulted in Table 2.4.

### ■ EXERCISE 2.9

To understand stability regions better:

- show that the stability region of the backward Euler method includes the left half of the complex plane; and
- show that the stability region of the trapezoidal rule is the left half of the complex plane.

- Heun's method can be viewed as a predictor–corrector formula resulting from a prediction with Euler's method and a single correction with the trapezoidal rule. Show that the stability region of Heun's method is finite.

## ■ EXERCISE 2.10

To appreciate better the mechanics of the various kinds of methods, write simple MATLAB programs to solve

$$y' = \begin{pmatrix} 0 & 1 \\ -1 & 0 \end{pmatrix} y, \quad y(0) = \begin{pmatrix} 0 \\ 1 \end{pmatrix}$$

on $[0, 1]$. Use a constant step size of $h = 0.1$ and plot the numerical solution together with the analytical solution.

- Solve the IVP with Heun's method. For this write a function of the form `[tnp1, ynp1]` `= Heun(tn,yn,h,f)`. This function is to accept as input the current solution $(t_n, y_n)$, the step size $h$, and the handle $f$ of the function for evaluating the ODEs. It advances the integration to $(t_{n+1}, y_{n+1})$. You may not be familiar with the way MATLAB evaluates functions that have been passed to another function as an input argument. This is done using the built-in function `feval` as illustrated by a function for taking a step with Euler's method:

```
function [tnp1,ynp1] = Euler(tn,yn,h,f)
yp = feval(f,tn,yn);
ynp1 = yn + h*yp;
tnp1 = tn + h;
```

See the MATLAB documentation for more details about `feval`.
- Solve the IVP with the trapezoidal rule evaluated with simple iteration. Heun's method can be viewed as a predictor–corrector pair resulting from a prediction with Euler's method and a single correction with the trapezoidal rule. Modify the function `Heun` to obtain a function `AM2si` that evaluates the trapezoidal rule by iterating to completion. It is not easy to decide when to stop iterating. For this exercise, if `p` is the current iterate and the new iterate is `c`, accept the new iterate if `norm(c - p)` `< 1e-3*norm(c)` and otherwise continue iterating. If your iteration does not converge in ten iterations, terminate the run with an error message.
- This IVP is not stiff, but solve it with the trapezoidal rule to see what is involved in the solution of stiff IVPs. Modify the function `AM2si` so that it has the form `[tnp1,ynp1]` `= AM2ni(tn,yn,h,f,dfdy)`. Here `dfdy` is the handle of a function $df\,dy(t, y)$ for evaluating the Jacobian $\frac{\partial f}{\partial y}(t, y)$. On entry to `AM2ni`, use `feval` to evaluate the (constant) Jacobian $J$, form the iteration matrix $M = I - 0.5hJ$, and compute its $LU$ decomposition. Use this factorization to compute the iterates. As explained in the text, it is important to code the iteration so that

you compute the correction $\Delta_m$ and then the new iterate $c$ as the sum of the current iterate $p$ and the correction. Much as with simple iteration, accept the new iterate if $\|\Delta_m\|$ is less than `1e-3*norm(c)` and otherwise continue iterating. If your iteration does not converge in ten iterations, terminate the run with an error message. (Newton's method converges immediately for this ODE, but you are to code for general ODEs so as to see what is involved.)

### ■ EXERCISE 2.11

The text discusses the solution of the stiff IVP

$$y' = -100y + 10, \quad y(0) = 1$$

on $0 \le t \le 10$ with AB1. It is said that a small step size is necessary in the beginning to resolve the solution in a boundary layer, but after a while the solution is nearly constant and a large step size can be used. Justify this statement by finding the largest step size $h$ for which the magnitude of the leading term in the local truncation error at $t_n$ is no greater than an absolute error tolerance of $\tau$.

### ■ EXERCISE 2.12

For simplicity and convenience, applications codes often use a constant step size when integrating with the backward Euler formula. This formula has very good stability properties, but if misused these properties can lead to a numerical solution that is qualitatively wrong. As a simple example, integrate the IVP

$$y' = 10y, \quad y(0) = 1$$

with the backward Euler method and a step size $h = 1$. How does the numerical solution compare to the analytical solution $y(t) = e^{10t}$? What's going on?

### ■ EXERCISE 2.13

A difficulty related to that of Exercise 2.12 is illustrated by the IVP

$$y' = -\frac{1}{t^2} + 10\left(y - \frac{1}{t}\right), \quad y(1) = 1$$

If we expect on physical grounds that the solution decays to zero, we might be content to approximate it using the backward Euler method and a constant step size. The analytical solution of this problem is $y(t) = t^{-1}$, so it does decay. The results of solving the IVP with step size $h = 1$, displayed in Table 2.6, appear to be satisfactory.

Now solve this IVP with each of the variable–step-size codes `ode45` and `ode15s`. The numerical solutions are terrible. These are quality solvers, so what is going on? What do these numerical solutions tell you about this IVP?

Table 2.6: *Backward Euler solution.*

| $t$ | $y$ |
|------|-------|
| 2.0 | 0.472 |
| 3.0 | 0.330 |
| 4.0 | 0.248 |
| 5.0 | 0.199 |
| 6.0 | 0.166 |
| 7.0 | 0.142 |
| 8.0 | 0.124 |
| 9.0 | 0.110 |
| 10.0 | 0.099 |

The difficulty illustrated by this exercise may not be so obvious when it arises in practice. Section 2.3.3 discusses how the solution of a PDE might be approximated by the solution of a system of ODEs. There is some art in this. When using finite differences to approximate some kinds of PDEs, it may happen that an "obvious" approximating system of ODEs is unstable. It is possible to damp instabilities by using the backward Euler method and a "large" step size, but whether the numerical solution will then faithfully model the solution of the PDE is problematical.

### ■ EXERCISE 2.14

The backward Euler method is misused in another way in some popular applications codes. It is recognized that the stability of an implicit method is needed, but – in order to keep down the cost – the formula is evaluated by a predictor–corrector process with only one correction. You can't have it both ways: If you want the stability of an implicit method, you must go to the expense of evaluating it properly. To see an example of this, show that the stability region of the predictor–corrector pair consisting of a prediction with the forward Euler method and one correction with the backward Euler method is finite. The qualitative effect of a single correction is made clear by showing that the predictor–corrector pair is not stable for (say) $z = h\lambda = -2$, whereas the backward Euler method is stable for all $z$ in the left half of the complex plane.

### ■ EXERCISE 2.15

This exercise will help you understand stiffness in both theory and practice. O'Malley (1991) models the concentration of a reactant $y(t)$ in a combustion process with the ODE

$$y' = f(y) = y^2(1 - y)$$

that is to be integrated over $[0, 2\varepsilon^{-1}]$ with initial condition $y(0) = \varepsilon$. He uses perturbation methods to analyze the behavior of the solution for small disturbances $\varepsilon > 0$ from the pre-ignition state. The analytical work is illuminating but, for the sake of simplicity, just solve the IVP numerically for $\varepsilon = 10^{-4}$ with the solver ode15s based on the BDFs and the program

```
function ignition
epsilon = 1e-4;
options = odeset('Stats','on');
ode15s(@ode,[0, 2/epsilon],epsilon,options);


%=====================================
function dydt = ode(t,y)
dydt = y^2*(1 - y);
```

This program displays the solution to the screen as it is computed and displays some statistics at the end of the run. You will find that the solution increases slowly from its initial value of $\varepsilon$. At a time that is $O(\varepsilon^{-1})$, the reactant ignites and increases rapidly to a value near 1. This increase takes place in an interval that is $O(1)$. For the remainder of the interval of integration, the solution is very near to its limit of 1. Modify the program to solve the IVP with the solver ode45 based on an explicit Runge–Kutta pair: all you must do is change the name of the solver. You will find that the numerical integration stalls after ignition, despite the fact that the solution is very nearly constant then. To quantify the difference, use tic and toc to measure the run times of the two solvers over the whole interval and over the first half of the interval. In this you should not display the solution as it is computed; that is, you should change the invocation of ode15s to

```
[t,y] = ode15s(@ode,[0, 2/epsilon],epsilon,options);
```

You will find that the nonstiff solver ode45 is rather faster on the first half of the interval because of the superior accuracy of its formulas – this despite the minimal linear algebra costs in ode15s due to the ODE having only one unknown. You will find that the stiff solver ode15s is much faster on the whole interval because the explicit Runge–Kutta formulas of ode45 must use a small step size (to keep the integration stable in the last half of the interval) and the BDFs of ode15s do not. The statistics show that the Runge–Kutta code has many failed steps in the second half of the integration. This is typical when solving a stiff problem with a method that has a finite stability region.

To understand the numerical results, work out the Jacobian $\frac{\partial f}{\partial y}$. Because there is a single equation, the only eigenvalue is the Jacobian itself. Work out a Lipschitz constant for $0 \leq y \leq 1$. You will find that it is not large, so the IVP can be stiff only on long intervals. In the first part of the integration the solution is positive, slowly varying, and $O(\varepsilon)$. Use the sign of the eigenvalue to argue that the IVP is unstable in a linear stability analysis and hence is not stiff in this part of the integration. Argue that it is only moderately unstable on this interval of length $O(\varepsilon^{-1})$, so we can expect to solve it accurately. The change in the solution at ignition is quite sharp when plotted on $[0, 2\varepsilon^{-1}]$, but by using `zoom` you will find that most of the change occurs in $[9650, 9680]$. Argue that in an interval of size $O(1)$ like this, the IVP is not stiff. In the last half of the interval of integration, the solution is close to 1 and slowly varying. Use the sign of the eigenvalue to argue that the IVP is stable then, and argue that the IVP is stiff on this interval of length $O(\varepsilon^{-1})$.

### ■ EXERCISE 2.16

The boundary of the stability region for a linear multistep method can be computed by the *root locus method*. An LMM applied to the test equation $y' = \lambda y$ with constant step size $h$ has the form

$$\sum_{i=0}^{k} \alpha_i y_{n+1-i} - h \sum_{i=0}^{k} \beta_i (\lambda y_{n+1-i}) = \sum_{i=0}^{k} (\alpha_i - h\lambda\beta_i) y_{n+1-i} = 0$$

This is a linear difference equation of order $k$ with constant coefficients that can be studied much like a linear ODE of order $k$ with constant coefficients. Each root $r$ of the characteristic polynomial

$$\sum_{i=0}^{k} (\alpha_i - h\lambda\beta_i) r^{k-i}$$

provides a solution of the difference equation of the form $y_m = r^m$ for $m = 0, 1, 2, \ldots$. As with ODEs, there are other kinds of solutions when $r$ is a multiple root; but also as with ODEs it is found that, for a given $z = h\lambda$, all solutions of the difference equation are bounded and thus the linear multistep method is stable if all the roots of the characteristic polynomial have magnitude less than 1. For a point $z$ on the boundary of the stability region, a root $r$ has magnitude equal to 1. We can use this observation to trace the boundary of the stability region by plotting all the $z$ for which $r$ is a root of magnitude 1. For this it is convenient to write the polynomial as $\rho(r) - z\sigma(r)$, where

$$\rho(r) = \sum_{i=0}^{k} \alpha_i r^{k-i}, \qquad \sigma(r) = \sum_{i=0}^{k} \beta_i r^{k-i}$$

With these definitions, the boundary of the stability region is the curve

$$z = \frac{\rho(r)}{\sigma(r)}$$

for $r = e^{i\theta}$ as $\theta$ ranges from 0 to $2\pi$. Plot the boundary of the stability region for several of the linear multistep methods discussed in the text – for example, the forward Euler method (AB1), the backward Euler method (AM1, BDF1), the trapezoidal rule (AM2), and BDF2. Some other linear multistep formulas that you might try are AM3,

$$y_{n+1} = y_n + h\left[\frac{5}{12}y'_{n+1} + \frac{8}{12}y'_n - \frac{1}{12}y'_{n-1}\right]$$

and BDF3,

$$y_{n+1} = \frac{18}{11}y_n - \frac{9}{11}y_{n-1} + \frac{2}{11}y_{n-2} + \frac{6}{11}h'y_{n+1}$$

The root locus method gives you only the boundary of the stability region. The BDFs have stability regions that are unbounded, so they are stable *outside* the curve you plot. The trapezoidal rule is a little different because, as Exercise 2.9 asks you to prove directly, it is stable in the left half of the complex plane. The other formulas mentioned have finite stability regions and so are stable inside the curve.

# 2.3 Solving IVPs in MATLAB

In the simplest use of the MATLAB solvers, all you must do is tell the solver what IVP is to be solved. That is, you must provide a function that evaluates $f(t, y)$, the interval of integration, and the vector of initial conditions. MATLAB has a good many solvers implementing diverse methods, but they can all used in *exactly* the same way. The documentation suggests that you try `ode45` first unless you suspect that the problem is stiff, in which case you should try `ode15s`. In preceding sections we have discussed what stiffness is, but the practical definition is that if `ode15s` solves the IVP *much* faster than `ode45` then the problem is stiff. Many people have the impression that if an IVP is hard for a code like `ode45` then it must be stiff. No, it might be hard for a different reason and so be hard for `ode15s`, too. On the other hand, if a problem is hard for `ode45` and easy for `ode15s`, then you almost certainly have a stiff problem.

There is a programming issue that deserves comment. MATLAB programs can be written as script files or functions. We have preferred to write the examples as functions because then auxiliary functions can be provided as subfunctions. Certainly it is convenient to have the function defining the ODEs available in the same file as the main program. This becomes more important when several functions must be supplied, as when solving IVPs with complications such as event location. Supplying multiple auxiliary functions is always necessary when solving BVPs and DDEs.

### EXAMPLE 2.3.1

In Chapter 1 we looked at the analytical solution of a family of simple equations. (Exercise 2.17 has you solve them numerically.) Among the problems was

$$y' = y^2 + t^2$$

We found the general solution to be a rather complicated expression involving fractional Bessel functions. Still, the equation is simple enough that you can solve analytically for the solution that has $y(0) = 0$ and plot it for $0 \leq t \leq 1$ at the command line with

```
>> y = dsolve('Dy = y^2 + t^2','y(0) = 0')
>> ezplot(y,[0,1])
```

This way of supplying the ODEs is satisfactory for very simple problems, but the numerical solvers are expected to deal with large and complicated systems of equations. Accordingly, they expect the ODEs to be evaluated with a (sub)function. For this problem, the ODE might be coded as

```
function dydt = f(t,y)
dydt = y^2 + t^2;
```

and saved in the file f.m. The IVP is solved numerically on [0, 1] and the solution is plotted by the commands

```
[t,y] = ode45(@f,[0,1],0);
plot(t,y)
```

which you can code either as a script file or function. Of course, with only two commands, you might as well enter them at the command line. The first input argument tells ode45 which function is to be used for evaluating the ODEs. This is done with a function handle, here @f. The second input argument is the interval of integration [a, b] and the third is the initial value. The solver computes approximations $y_n \approx y(t_n)$ on a mesh

$$a = t_0 < t_1 < \cdots < t_N = b$$

that is chosen by the solver. The mesh points are returned in the array t and the corresponding approximate solutions in the array y. This output is in a form convenient for plotting. Figure 2.1 shows the result of this computation. When run as a script (or from the command line) you have available the mesh t and the solution y on this mesh, so you can, for example, look at individual components of systems or plot with logarithmic scales.

All the input arguments are required even to define the IVP. Indeed, the only argument that might be questioned is the interval of integration. However, we have seen that the stability of an IVP is fundamental to its numerical solution and this depends on both the length of the interval and the direction of integration. The MATLAB problem-solving environment and the design of the solvers make it possible to avoid the long call lists of

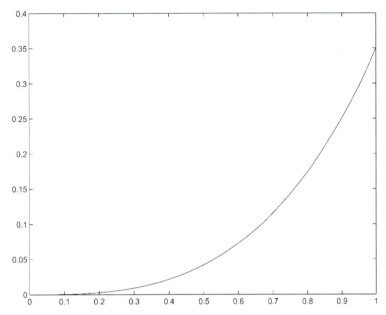

Figure 2.1: Solution of the ODE $y' = y^2 + t^2$ with initial condition $y(0) = 0$.

arguments that are typical of solvers written for compiled computation in general scientific computing. For instance, MATLAB makes it possible to avoid the tedious and error-prone specification of storage and data types that are necessary for compiled computation. This is accomplished by making heavy use of default values for optional arguments.

The MATLAB PSE makes convenient the output of solution arrays of a size that cannot be determined in advance and provides convenient tools for plotting these arrays of data. The typical solver for general scientific computing has you supply all the $t_n$ where you want answers. Storage issues are an important reason for this, but of course it may be that you actually want answers at specific points. One way that the MATLAB IVP solvers provide this capability is by overloading the argument that specifies the interval. If you supply an array [t0 t1 ... tf] of more than two entries for the interval of integration, the solver interprets this as an instruction to return approximations to the solution values $y(t0), y(t1), \ldots, y(tf)$. When called in this way, the output array t is the same as the input array. That the number and placement of specified output points has little effect on the computation is of great practical importance. The output points must be in order, that is, either $a = t0 < t1 < \cdots < tf = b$ or $a = t0 > t1 > \cdots > tf = b$. As a concrete example, suppose that we had wanted to construct a table of the solution of the last example at ten equally spaced points in the interval $[0, 1]$. This could be achieved with the script

```
tout = linspace(0,1,10);
[t,y] = ode45(@f,tout,0);
```

There is another way to receive output from the IVP solvers that has some advantages. The output can be in the form of a structure that can be given any name, but let us suppose that it is called `sol`. For the present example the syntax is

```
sol = ode45(@f,[0,1],0);
```

In this way of using the solvers, the usual output `[t,y]` is available as fields of the solution structure `sol`. Specifically, the mesh t is returned as the field `sol.x` and the solution at these points is returned as the field `sol.y`. However, that is not the point of this form of output. The solvers actually produce a continuous solution $S(t)$ on all of $[a, b]$. They return in the solution structure `sol` the information needed to evaluate $S(t)$. This is done with an auxiliary function `deval`. It has two input arguments, a solution structure and an array of points where you want approximate solutions. Rather than tell a solver to compute answers at `tout = linspace(0,1,10)`, you can have it return a solution structure `sol` and then compute answers at these points with the command

```
Stout = deval(sol,tout);
```

With this mode of output, you solve an IVP just once. Using the solution structure you can then compute answers anywhere you want. In effect, you have a function for the solution. Indeed, to make this more like a function, `deval` also accepts its arguments in the reverse order:

```
Stout = deval(tout,sol);
```

This mode of output is an option for the IVP solvers, but it is the only mode of output from the BVP and DDE solvers of MATLAB.

Some ODEs cannot be integrated past a critical point. How this difficulty is handled depends on the solver. It is easy to deal with the difficulty if the code produces output at a specific point by stepping to that point, because you simply do not ask for output beyond the critical point. However, this is an inefficient design. Modern solvers integrate with the largest step sizes they can and then evaluate continuous extensions to approximate the solution at the output points. Accordingly, they must be allowed to step past output points. So, what do you do when the solver cannot step past some point? Many codes in general scientific computing have an option for you to inform them of this relatively unusual situation. The MATLAB solvers handle this in a different way: they accept only problems set on an interval. They do not integrate past the end of this interval, nor do they evaluate the ODE at points outside the interval. Accordingly, for this example it does not matter whether the ODE is even defined for $t > 1$.

## EXAMPLE 2.3.2

Example 2.3.1 is unusual in that the IVP is solved from the command line. The program ch2ex1.m is more typical. It solves the ODEs

$$x_1' = -k_1 x_1 + k_2 y$$
$$x_2' = -k_4 x_2 + k_3 y$$
$$y' = k_1 x_1 + k_4 x_2 - (k_1 + k_3) y$$

with initial values

$$x_1(0) = 0, \quad x_2(0) = 1, \quad y(0) = 0$$

on the interval $0 \le t \le 8 \cdot 10^5$. The coefficients here are

$$k_1 = 8.4303270 \cdot 10^{-10}, \quad k_2 = 2.9002673 \cdot 10^{11},$$
$$k_3 = 2.4603642 \cdot 10^{10}, \quad k_4 = 8.7600580 \cdot 10^{-6}$$

By writing the main program as a function, we can code the evaluation of the ODEs as the subfunction odes. The output of this subfunction must be a column vector, which is achieved here by initializing dydt as a column vector of zeros. The approximations to $y_i(t_n)$ for $n = 1, 2, \ldots$ are returned as y(:,i). The program uses this output to plot $x_1(t)$ and $x_2(t)$ together with a linear scale for $t$ and to plot $y(t)$ separately with a logarithmic scale for $t$.

This is the proton transfer problem discussed in Section 1.4, where the two plots are provided as Figures 1.6 and 1.7. We pointed out that the default absolute error tolerance of $10^{-6}$ applied to all solution components is inappropriate for $y(t)$ because Figure 1.7 shows that it has a magnitude less than $3 \cdot 10^{-17}$. All optional specifications to the MATLAB IVP solvers are provided by means of the auxiliary function odeset and *keywords*. The command help odeset provides short explanations of the various options and the command odeset by itself provides reminders. Error tolerances are the most common optional input. An absolute error tolerance of $10^{-20}$ is more appropriate for this IVP. When the absolute error tolerance is given a scalar value, it is applied to all solution components. When it is given a vector value, the entries in the vector of absolute error tolerances are applied to corresponding entries in the solution vector. Any option not specified is given its default value. Here we use the default value of $10^{-3}$ for the relative error tolerance (which is always a scalar), but if we had wanted to assign it a value of (say) $10^{-4}$ then we would have used the command

```
options = odeset('AbsTol',1e-20,'RelTol',1e-4)
```

Options can be assigned in any order and the keywords are not case-sensitive. After forming an options structure with odeset, it is provided as the optional fourth input argument of the solver. Here `options` is a natural name for the options structure, but you can use whatever name you like.

```
function ch2ex1
% x_1 = y(1), x_2 = y(2), y = y(3)

options = odeset('AbsTol',1e-20);
[t,y] = ode15s(@odes,[0 8e5],[0; 1; 0],options);
plot(t,y(:,1:2))
figure
semilogx(t,y(:,3))


%====================================================
function dydt = odes(t,y)
k = [8.4303270e-10 2.9002673e+11 2.4603642e+10 8.7600580e-06];
dydt = zeros(3,1);
dydt(1) =  -k(1)*y(1) + k(2)*y(3);
dydt(2) =  -k(4)*y(2) + k(3)*y(3);
dydt(3) =   k(1)*y(1) + k(4)*y(2) - (k(2) + k(3))*y(3);
```

We solved this problem with the BDF code `ode15s` because the IVP is stiff. In simplest use, all the IVP solvers are interchangeable. To use the Runge–Kutta code `ode45` instead, all you must do is change the name of the solver in `ch2ex1.m`. You are asked to do this in Exercise 2.18 so that you can see for yourself why special methods are needed for stiff problems.

### EXAMPLE 2.3.3

By default, the solvers return the solution at all steps in the course of the integration. To get more control over the output, you can write an *output function* that the solver will call at each step with the solution it has just computed. Several output functions are supplied with the solvers. As a convenience, if you do not specify any output arguments then the solver understands that you want to use the output function `odeplot`. This output function displays all of the solution components as they are computed. Exercises 2.15 and 2.18 are examples. Often you do not want to see all of the components. You can control which components are displayed with the `OutputSel` option. Other output functions make it convenient to plot in a phase plane. Exercise 2.21 has you experiment with one.

Along with the example programs of the text is a program `dfs.m` that provides a modest capability for computing and plotting solutions of a scalar ODE, $y' = f(t, y)$. Exercise 1.4 discusses how to use the program, but we consider it here only to show how to write an output function. Some solvers allow users to specify a minimum step size, *hmin*, and terminate the integration if a step size this small appears to be necessary. The MATLAB solvers do not provide for a minimum step size, but dfs.m shows how to obtain this capability with an output function. Relevant portions of the program are

```
function dfs(fun,window,npts)
    .
    .
    .
hmin = 1e-4*(wR - wL);
    .
    .
    .
options = odeset('Events',@events,'OutputFcn',@outfcn,...
    .
    .
    .
[t,y] = ode45(@F,[t0,wR],y0,options);
plot(t,y,'r')
[t,y] = ode45(@F,[t0,wL],y0,options);
plot(t,y,'r')
    .
    .
    .
function status = outfcn(t,y,flag)
global hmin
persistent previoust
status = 0;
switch flag
case 'init'
    previoust = t(1);
case ''
    h = t(end) - previoust;
    previoust = t(end);
    status = (abs(h) <= hmin);
case 'done'
    clear previoust
end
```

The option OutputFcn communicates to the solver the name of the output function, here outfcn. When there is an output function present, the solver calls it at each step with arguments t,y,flag. The arguments t and y are values of the independent and dependent variables, and the string flag indicates the circumstances of the call. The output function returns a variable `status`. If you want to continue the integration, return a value of 0 (false); if you want to terminate the run, return a value of 1 (true). The solver calls first with `flag` equal to `'init'` so that the output function can initialize itself. For example, if you wanted to write output to a file, you could open the file at this time. In this first call the solver provides the output function with the interval of integration $[a, b]$ in `t` and $y(a)$ in `y`. In the present example a variable `previoust` is initialized to the initial value of the independent variable, `t(1)`. It is declared to be `persistent` so that it will be retained between calls. After each step of the integration, the solver calls the output function with `flag` equal to `' '`. Generally `t` is the value of the independent variable at the end of the step and `y` is the approximate solution there, but a complication arises in this example because the integration is performed using `ode45`. The solvers have an option called `Refine` that can be given an integer value to have the solver return that many approximate solutions at equally spaced points in the span of the step. The default value of `Refine` is 1 for all the solvers except `ode45`, for which it is 4. You can do whatever you like with these approximate solutions in the output function. You might, for example, write `t` and selected components of `y` to a file. When `Refine` is larger than 1, the approximate solutions appear in order, so we can use `t(end)` to obtain the value of the independent variable at the end of the step that we need to compute the current step size in this example. If the step size is not larger than the minimum step size, `status` is set to 1 so that the integration will be terminated. Exercise 2.29 asks you to modify the output function of `dfs.m` so as to terminate the run in other circumstances as well. At the end of the run, the solver calls the output function with `flag` equal to `'done'` so that the output function can finish up. For example, it might close an output file. Here `previoust` is cleared just to illustrate the case.

### ■ EXERCISE 2.17

In Chapter 1 it was asserted that there is no essential difference solving numerically IVPs for the equations

- $y' = y^2 + 1$
- $y' = y^2 + t$
- $y' = y^2 + t^2$

See for yourself by solving the equations with, say, `ode45` on the interval $[0, 1]$ with initial value $y(0) = 0$ and then plotting the solutions.

### ■ EXERCISE 2.18

Modify `ch2ex1.m` so that the approximate solutions are displayed as they are computed. As explained in Example 2.3.3, this is what happens when you do not specify any output arguments. You will find that `ode15s` solves the IVP easily. Although it appears that `ode45` is stuck at the initial point, it *is* solving the problem. This will be clear when you click the "Stop" button after you tire of waiting for the solver to finish up. At that time the numerical solution will be plotted on a scale that shows how far the integration has progressed. You will find that `ode45` is solving the IVP, but it is advancing the integration *very* slowly.

### ■ EXERCISE 2.19

Before effective codes for stiff IVPs were widely available, some stiff problems were solved by singular perturbation methods, perhaps in combination with numerical methods for nonstiff IVPs. In this way Lapidus et al. (1973) solve the following model of the thermal decomposition of ozone:

$$\frac{dx}{dt} = -x - xy + \varepsilon \kappa y$$

$$\varepsilon \frac{dy}{dt} = x - xy - \varepsilon \kappa y$$

Here $x$ is the reduced ozone concentration, $y$ is the reduced oxygen concentration, and the parameters $\varepsilon = 1/98$ and $\kappa = 3$. (Notice that $y'(t)$ is multiplied by the small parameter $\varepsilon$.) The IVP is to be solved on $[0, 240]$ with initial conditions $x(0) = 1$ and $y(0) = 0$. Lapidus and colleagues observe that singular perturbation methods are not very accurate for this IVP. Ironically, this is because $\varepsilon$ is not very small, which is to say that the IVP is not very stiff. Nowadays numerical solution of this IVP is routine. Solve it with `ode15s` and plot $y(t)$ with `semilogx` and `axis([0.01 100 0 1])` for comparison with Figure 3 of Lapidus et al. (1973). Verify that the IVP is not very stiff by showing that you can solve the IVP easily with `ode45`. Using `odeset`, set the option `Stats` to on for the two runs to compute statistics that will help with this verification.

### ■ EXERCISE 2.20

The text states that, when evaluating an implicit method, the iteration matrix is ill-conditioned if the IVP is stiff. See for yourself by modifying `ch2ex1.m` so as to display a condition number for each of the iteration matrices formed in the course of the integration. For this you will need to copy `ode15s.m` to your working directory and rename it to, say, `mode15s.m`. The source code for `ode15s` is found in the subdirectory `/toolbox/matlab/funfun/` of your installation directory for MATLAB. You will also need copies of some auxiliary functions that `ode15s` calls, namely `ntrp15s.m`, `odearguments.m`, `odeevents.m`, `odejacobian.m`, and `odemass.m`. They are found in `/toolbox/matlab/funfun/private/`. To estimate a condition number for the iteration matrix and communicate it to the main program, search `mode15s.m` for the command

```
[L,U] = lu(Miter);
```

which uses the MATLAB function `lu` to compute an *LU* factorization. It occurs twice, once as the solver initializes and once in the main loop. The first time you might follow this command with

```
global tcond cond
tcond = t;
cond = condest(Miter);
```

This initializes a pair of output arrays that keep track of where the iteration matrix is formed, computes an estimate of its condition in the 1-norm using the MATLAB function `condest`, and makes this information available outside the solver. You might then follow the second appearance of the command by

```
tcond = [tcond t];
cond = [cond condest(Miter)];
```

to extend the output arrays each time an iteration matrix is formed and factored. After this preparation it is easy to monitor a condition number for the iteration matrix as you solve an IVP with `mode15s`. To do this with `ch2ex1.m`, all you must do is add

```
global tcond cond
```

to gain access to the information and follow the computations with a plot such as

```
figure
loglog(tcond,cond,'*-')
```

You will find that, in the first part of the integration, the condition number is comparable to 1, the smallest possible value. It then grows steadily to values comparable to 1/eps, the largest possible value. The condition number is reduced somewhat for the last step. Using the appearance of the solution components as plotted by `ch2ex1.m` and a plot of the step sizes,

```
figure
loglog(t(1:end-1),diff(t))
```

explain why you might have expected the condition number to behave as described.

■ **EXERCISE 2.21**

Two of the output functions that accompany the MATLAB IVP solvers, `odephas2` and `odephas3`, plot solutions in a phase space as they are computed. Solutions of the ODEs

$$y'_1 = -y_2 - \frac{y_1 y_3}{r}$$

$$y'_2 = y_1 - \frac{y_2 y_3}{r}$$

$$y'_3 = \frac{y_1}{r}$$

where

$$r = \sqrt{y_1^2 + y_2^2}$$

lie on a torus in phase space. Using `ode45`, solve these equations on the interval $0 \leq t \leq 10$ with initial values

$$(y_1(0), y_2(0), y_3(0)) = (3, 0, 0)$$

and then plot the solution in 3-dimensional phase space as it is computed by setting `OutputFcn` to `@odephas3`. The plot routines draw straight lines between successive output points. When the step sizes are "large", the output points may be so far apart that the graph is not smooth. This is more likely to happen when plotting in phase space. If it happens, you can use the option `Refine` to have the solver compute additional output points in the span of each step by evaluating a continuous extension. To see this effect clearly, compute two figures. For one figure set `Refine` to 1 so that there is one output point per step. For `ode45` the default value of `Refine` is 4, but this graph would benefit from more output points, so set `Refine` to 10 for the second figure. The auxiliary function `odeset` allows you to alter options as well as set them: If you compute the first figure with

```
options = odeset('OutputFcn',@odephas3,'Refine',1);
```

then you can reset the value of `Refine` for the second figure with

```
options = odeset(options,'Refine',10);
```

### ■ EXERCISE 2.22

Raghothama & Narayanan (2002) consider the effects of parametric excitation for the ODE

$$x''(t) = -2\zeta x'(t) - x - 1.5x^2(t) - 0.5x^3(t) + f_2 \cos(\Omega_2 t) + f_3 \cos(\Omega_3 t)$$

where $f_2 = 0.05$, $\Omega_2 = 1$, $\Omega_3 = 2$, $\zeta = 0.1$, and $f_3$ is treated as a bifurcation parameter. The authors demonstrate that, as $f_3$ is changed, the phase curves $(x(t), x'(t))$ can be qualitatively different. Use `ode45` to solve the ODE for the following four cases corresponding to the cases in Figure 9 of Raghothama & Narayanan (2002). Integrate with `RelTol` = `1e-8` and `AbsTol` = `1e-8` from $t = 0$ to $t = 500$. To show the limiting behavior, just plot the phase curve for $400 \leq t \leq 500$. If you have the integration coded as [t,x] = ode45..., you can do this by first finding the indices of mesh points in this interval with `ndx = find(t >= 400)` and then plotting with `plot(x(ndx,1),x(ndx,2))`. Your curves should represent periodic motion of period 1, 2, 4, and 8, respectively.

- $x(0) = 0$, $x'(0) = -0.60$, $f_3 = 0.50$
- $x(0) = 0$, $x'(0) = -0.80$, $f_3 = 0.92$
- $x(0) = 0$, $x'(0) = -0.59$, $f_3 = 0.99$
- $x(0) = 0$, $x'(0) = -0.80$, $f_3 = 1.005$

The authors further demonstrate that chaotic behavior can occur. Solve the IVP in the same way with $x(0) = 0$, $x'(0) = 0$, and $f_3 = 1.35$. Your phase curve should resemble a (right-handed) catcher's mitt.

## 2.3.1 Event Location

Along with the example programs of the text is a program `dfs.m` that provides a modest capability for computing and plotting solutions of a scalar ODE, $y' = f(t, y)$. Exercise 1.4 discusses how to use this program, but only a few details are needed here. Solutions are plotted in a window that you specify by an array $[wL, wR, wB, wT]$. When you click on a point $(t_0, y_0)$, the program computes the solution $y(t)$ that has the initial value $y(t_0) = y_0$ and plots it as long as $wL \leq t \leq wR$ and $wB \leq y \leq wT$. By calling `ode45` with interval $[t_0, wR]$, the program computes $y(t)$ from the initial point to the right edge of the plot window. It then calls the solver with the interval $[t_0, wL]$ to compute $y(t)$ from the initial point to the left edge. But what if the solution goes out the bottom or top of the plot window along the way? What we need is the capability of terminating the integration if there is a $t^*$ where either $y(t^*) = wB$ or $y(t^*) = wT$.

While approximating the solution of an IVP, we sometimes need to locate points $t^*$ where certain *event functions*

$$g_1(t, y(t)), g_2(t, y(t)), \ldots, g_k(t, y(t))$$

vanish. Determining these points is called *event location*. Sometimes we just want to know the solution, $y(t^*)$, at the time, $t^*$, of the event. On other occasions we need to terminate the integration at $t^*$ and possibly start solving a new IVP with initial values and ODEs that depend on $t^*$ and $y(t^*)$. Sometimes it matters whether an event function is decreasing or increasing at the time of an event. In the `dfs.m` program, there are two event functions ($wB - y$ and $wT - y$), both events are terminal, and we don't care whether the event function is increasing or decreasing at an event. All the MATLAB IVP solvers have a powerful event location capability. Most continuous simulation languages and a few of the IVP solvers widely used in general scientific computing have some kind of capability of locating events. Event location can be ill-posed and some of the algorithms in use are crude. The difficulties of this task are often not appreciated, so one of the aims of this section is to instill some caution when using an event location capability. A deeper discussion of the issues can be found in Shampine, Gladwell, & Brankin (1991) and Shampine & Thompson (2000).

Most codes that locate events monitor the event functions for a change of sign in the span of a step. If, say, $g_i(t_n, y_n)$ and $g_i(t_{n+1}, y_{n+1})$ have opposite signs, then the solver uses standard numerical methods to find a root of the algebraic equation $g_i(t, y(t)) = 0$ in $[t_n, t_{n+1}]$. In finding a root, it is necessary to evaluate the event function $g_i$ at a number of points $t$. Here is where a continuous extension is all but indispensable. Without a continuous extension, for each value of $t$ in the interval $[t_n, t_{n+1}]$ where we must evaluate $g_i$, we must take a step with the ODE solver from $t_n$ to $t$ in order to compute the approximation to $y(t)$ that we need in evaluating $g_i(t, y(t))$. It is much more efficient simply to evaluate a polynomial approximation $S(t)$ to $y(t)$ that is accurate for all of the interval $[t_n, t_{n+1}]$. Locating the event then amounts to computing zeros of the equation $g_i(t, S(t)) = 0$. Often there are a number of event functions. They must be processed simultaneously because it is possible that, in the course of locating a root of one event function, we discover that another event function has a root and it is closer to $t_n$. We must find the event closest to $t_n$ because the definition of the ODE might change there.

This outline of event location gives us good reasons for caution. The approach has no hope of finding events characterized by an even-order zero because there is no change of sign. It is entirely possible to miss a change of sign at an odd-order zero. That is because the step size is chosen to resolve changes in the solution, not changes in the event functions, so the code could step over several events and have no sign change at the end of the step. A related difficulty is that in applications we want the first root, the one closest to $t_n$, and in general there is no way to be sure we compute it. Some of the event functions that arise in practice are not smooth, which makes the computation of roots more difficult. Indeed, discontinuous event functions are not rare. As when computing roots of any algebraic equation, the root of an event function can be ill-conditioned, meaning that the value $t^*$ is poorly determined by the values of the event function. In root solving we generally assume that the function can be evaluated very accurately, but the situation is different here because we compute the argument $y(t)$ to only a specified accuracy. A fundamental issue, then, is how well $t^*$ is determined when we compute the function $y(t)$ to a specified accuracy. That must be considered when formulating the problem but it also has a consequence for the root solver, namely, how accurately should we have it compute $t^*$? Some codes ask users to supply tolerances for event location in addition to those for the integration. It is difficult to choose sensible values, so other codes (including those of MATLAB) take a different approach. They simply locate events about as accurately as possible in the precision of the computer. For this approach to be practical, the algorithm for computing roots must be fast in the usual case of smooth event functions. It must also be reasonably fast when an event function is not smooth. An article by Moler (1997) explains how the algorithm of MATLAB combines the robust and reasonably fast method of bisection with a scheme that is both fast for smooth event functions and vectorizable.

The examples that follow show that it is not hard to do event location with the MATLAB IVP solvers. Indeed, the only complication is that you must specify what kinds of events you want to locate and what you want the solver to do when it finds one. However, problems involving event location are often rather complicated because of what is done after an event occurs. For Example 2.3.4 this is just a matter of processing the events. Exercises 2.25 and 2.26 take up problems of this kind. Example 2.3.5 is complicated because a new IVP is formulated and solved after each event; Exercises 2.27 and 2.28 take up variants of this example. An example and a couple of exercises in later sections illustrate event location when there is something new about solving the ODEs: the ODEs of Example 2.3.7 are formulated in terms of a mass matrix. Exercise 2.32 is an extension of this example. The ODEs of Exercise 2.37 are singular at the initial point.

To see that event location does not need to be complicated, let us look at the relevant portions of dfs.m. There is a function `events` that evaluates the event functions and tells the solver what it is to do. A handle for this function is provided as the value of the option `Events`. The values of the two event functions for the input values `t,y` are returned by `events` in the output argument `value`. The vector `isterminal` tells the solver whether the events are terminal. Here both entries are ones, meaning that both are terminal events. The vector `direction` tells the solver if it matters whether the event function is increasing or decreasing at an event. Here both entries are zeros, meaning that the direction does not matter. The solver has some extra output arguments when there are events, but here we need only terminate the integration and so do not ask for this optional output.

```
function dfs(fun,window,npts)
  .
  .
  .
options = odeset('Events',@events,...
  .
  .
  .
[t,y] = ode45(@F,[t0,wR],y0,options);
plot(t,y,'r')
[t,y] = ode45(@F,[t0,wL],y0,options);
plot(t,y,'r')
  .
  .
  .
```

```
function [value,isterminal,direction] = events(t,y)
global wB wT
value = [wB; wT] - y;
isterminal = [1; 1];
direction = [0; 0];
```

It is nearly as easy to solve the event location problem of Exercise 2.23, but it is convenient then to use the solver's optional output arguments. This is just a sketch of event location; details are found in the examples that follow.

## EXAMPLE 2.3.4

Poincaré maps are important tools for the interpretation of dynamical systems. Very often they are plots of the values of the solution at a sequence of equally spaced times. These values are easily obtained by simply specifying an array tspan of these times. However, sometimes values are sought for which a linear combination of the solution values vanishes, and event location is needed to obtain these values. Koçak does this with the PHASER program in Lessons 13 and 14 of Koçak (1989). In his example, four ODEs for the variables $x_1(t)$, $x_2(t)$, $x_3(t)$, and $x_4(t)$ describing a pair of harmonic oscillators are integrated. The simple equations and initial conditions are found in the function ch2ex2.m. Plotting $x_1(t)$, $x_2(t)$, and $x_3(t)$ shows what appears to be a finite cylinder. As coded, you can click on the Rotate 3D tool and drag the figure to come to a good understanding of this projection of the solution. One view is provided here as Figure 2.2. The PHASER program can find and plot points where

$$Ax_1(t) + Bx_2(t) + Cx_3(t) + D = 0$$

Koçak computes two such Poincaré maps. One is a plot of the coordinates $(x_1(t^*), x_3(t^*))$ for values of $t^*$ such that $x_2(t^*) = 0$. The other is a plot of the coordinates $(x_1(t^*), x_2(t^*))$ for values of $t^*$ such that $x_3(t^*) = 0$. We accomplish this by employing two event functions, $g_1 = x_2$ and $g_2 = x_3$, that we evaluate in a subfunction here called events. As with other optional input, you tell the solver about this function by setting the option Events to the handle of this function. The event function must have as its input the arguments t,x and return as output a column vector value of the values of the event functions $g_1(t, x)$ and $g_2(t, x)$. We must write this function so that it also returns some information about what we want the code to do. For some problems we will want to terminate the integration when a certain event occurs. The second output argument of the event function is a column vector isterminal of zeros and ones. If we want to terminate the integration when $g_k(t, x) = 0$, we set isterminal(k) to 1 and otherwise we set it to 0. The initial point is a special case in this respect because sometimes an event that we want to regard as terminal occurs at the initial point. This special situation is

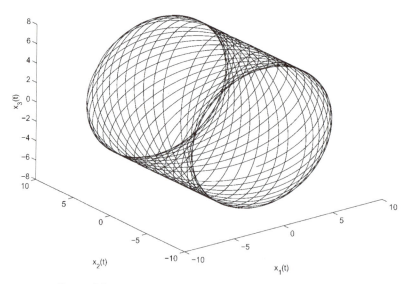

Figure 2.2: Lissajous figure for a pair of harmonic oscillators.

illustrated by Example 2.3.5 and Exercise 2.24. Because this situation is not unusual, the solvers treat any event at the initial point as *not* terminal. Sometimes it matters whether the event function is decreasing or increasing at an event. The third output argument of the event function is a column vector `direction`. If we are interested in events determined by the equation $g_k(t, x) = 0$ only at locations where the event function $g_k$ is decreasing, we set `direction(k)` to $-1$. If we are interested in events only when the function $g_k$ is increasing, we set it to $+1$. If we are interested in all events for this function, we set `direction(k)` to 0. For the current example, events are not terminal and we want to compute all events.

When we use the event location capability, the solver returns some additional quantities as exemplified here by `[t,x,te,xe,ie] = ode45...`. The array `te` contains the values of the independent variable where events occur and the array `xe` contains the solution there. The array `ie` contains an integer indicating which event function vanished at the corresponding entry of `te`. (If you prefer to have the output in the form of a solution structure `sol`, this information about events is returned in the fields `sol.xe`, `sol.ye`, and `sol.ie`, respectively.) If there are no events then all these arrays are empty, so a convenient way to check this possibility is to test `isempty(ie)`. One of the complications of event location is that, if there is more than one event function, we do not know in what order the various kinds of events will occur. This complication is resolved easily with the `find` function as illustrated by this example. For instance, the command

```
event1 = find(ie == 1);
```

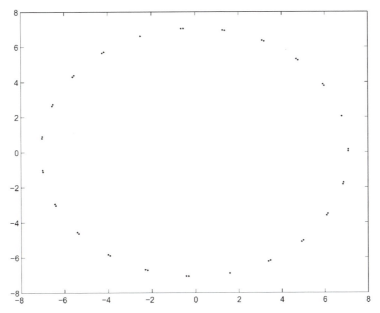

Figure 2.3: Poincaré map for a pair of harmonic oscillators.

returns in `event1` the indices that correspond to the first event function. They are then used to extract the corresponding solution values from `xe` for plotting. All the MATLAB IVP solvers have event location and they are all used in exactly the same way. Here we use `ode45`, and we use tolerances more stringent than the default values because we require an accurate solution over a good many periods. Figure 2.3 shows one of the two Poincaré maps computed by this program. The program also reports that

```
Event 1 occurred 43 times.
Event 2 occurred 65 times.
```

```
function ch2ex2
opts = odeset('Events',@events,'RelTol',1e-6,'AbsTol',1e-10);
[t,x,te,xe,ie] = ode45(@odes,[0 65],[5; 5; 5; 5],opts);
plot3(x(:,1),x(:,2),x(:,3));
xlabel('x_1(t)'), ylabel('x_2(t)'), zlabel('x_3(t)')
if isempty(ie)
    fprintf('There were no events.\n');
else
    event1 = find(ie == 1);
    if isempty(event1)
```

```
            fprintf('Event 1 did not occur.\n');
        else
            fprintf('Event 1 occurred %i times.\n',length(event1));
            figure
            plot(xe(event1,1),xe(event1,3),'*');
        end
        event2 = find(ie == 2);
        if isempty(event2)
            fprintf('Event 2 did not occur.\n');
        else
            fprintf('Event 2 occurred %i times.\n',length(event2));
            figure
            plot(xe(event2,1),xe(event2,2),'*');
        end
end

%===============================================================
function dxdt = odes(t,x)
a = 3.12121212;
b = 2.11111111;
dxdt = [a*x(3); b*x(4); -a*x(1); -b*x(2)];

function [value,isterminal,direction] = events(t,x)
value = [x(2); x(3)];
isterminal = [0; 0];
direction = [0; 0];
```

### EXAMPLE 2.3.5

An interesting problem that is representative of many event location problems models a ball bouncing down a ramp as in Figure 2.4. For simplicity we take the ramp to be the long side of the triangle with vertices $(0, 0)$, $(0, 1)$, and $(1, 0)$. At time $t$ the ball is located at the point $(x(t), y(t))$. We'll suppose that it is released from rest at a point above the end of the ramp; that is, we start from $x(0) = 0$, $y(0) > 1$, $x'(0) = 0$, and $y'(0) = 0$. In free fall the motion of the ball is described by the equations

$$x'' = 0, \quad y'' = -g$$

where the second equation represents the acceleration due to gravity and the gravitational constant is taken to be $g = 9.81$. We must write these equations as a system of first-order equations, which we do by introducing variables

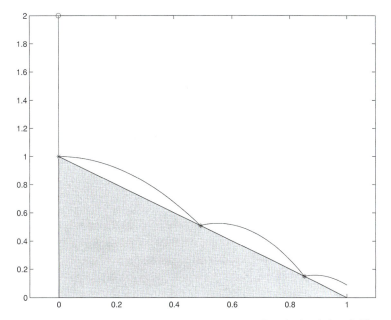

Figure 2.4: Path of the ball when the coefficient of restitution is $k = 0.35$.

$$y_1(t) = x(t), \quad y_2(t) = x'(t), \quad y_3(t) = y(t), \quad y_4(t) = y'(t)$$

These equations hold until the ball hits the ramp at a time $t^* > 0$. At this time the ball has moved to the right a distance $x(t^*)$ where the ramp has height $1 - x(t^*)$. (The first event has $x(t^*) = 0$.) The height of the ball is $y(t)$, so the event that the ball hits the ramp is

$$y(t^*) = 1 - x(t^*)$$

From this we see that the event function is

$$g_1 = y_3 - (1 - y_1)$$

This event is terminal because the model changes there. It is essential to locate the event in order to correctly model the subsequent trajectory of the ball, since there is nothing in the equations themselves to change the direction of the ball if a bounce time event is missed by the code.

After hitting the ramp, the ball rebounds and its subsequent motion is described by another IVP for the same ODEs. The initial position of the ball in the new integration is the same as the terminal position in the old. The effect of a bounce appears in the initial velocity of the new integration, which has a direction related to the velocity at $t^*$ by the geometry of the ramp and a magnitude reduced by a coefficient of restitution $k$, a constant with value $0 < k < 1$. We'll not go into the details, but the result is that the initial conditions for the new integration starting at time at $t^*$ are

$$(y_1(t^*), \; -ky_4(t^*), \; y_3(t^*), \; ky_2(t^*))$$

The same terminal event function is used to recognize the next time that the ball hits the ramp. However, at the initial point of this new IVP the ball is touching the ramp, so the terminal event function vanishes there. This shows why we want to treat the initial point as a special case for terminal events. There is another way to deal with this kind of difficulty that is convenient here. If we set `direction` to 0, the solvers report an event at the initial point in addition to the one that terminates the integration. We can avoid this by setting `direction` to $-1$. The distance of the ball from the ramp is increasing at the initial point, so with this setting the solver ignores the zero there and reports only the event of the ball dropping to the ramp. We repeat this computation of the path of the ball from bounce to bounce as it goes down the ramp.

The computation is complete when the ball reaches the end of the ramp, which is at $x(t^*) = 1$. To recognize this, we use the terminal event function $g_2 = y_1 - 1$. For some values of $k$ and some choices of initial height of the ball, the bounces cluster and the ball never reaches the end of the ramp. Our equations cannot model what actually happens then – the ball rolls down the ramp after it has finished bouncing. Hence, we must recognize that the bounces are clustering both to avoid clutter in the plot and to stop the integration. In the program this is done by quitting when the times at which successive bounces occur differ by less than 1%. There is a related difficulty when the ball bounces very nearly at the end of the ramp, so we also quit if it bounces at a point that is less than 1% of the distance from the end.

Some thought about the problem or some experimentation brings to our attention a difficulty of some generality – namely, that we do not know how long it will take for the ball to reach the end of the ramp, if ever. Certainly the terminal time depends on the initial height and on the coefficient of restitution, $k$. The solvers require us to specify an interval of integration, so we must guess a final time. If this guess isn't large enough, we'll need to continue integrating. Some IVP codes have a capability for *extending the interval,* but the MATLAB solvers do not. Instead, we solve a new IVP with initial data taken from the final data of the previous integration. This is an acceptable way to proceed because the MATLAB solvers start themselves efficiently and it won't be necessary to restart many times if our guesses are at all reasonable. Exercises 2.27 and 2.28 have you solve variants of the bouncing ball problem.

A couple of programming matters deserve comment. The path of the ball consists of several pieces, either because of bounces or because we had to extend the interval of integration. It is convenient to accumulate the whole path in a couple of arrays in a manner illustrated by the program. Because the ODEs are so easy to integrate, the solver does not take enough steps to result in a smooth graph. An array `tspan` is used to deal with this. In addition to plotting the solution with a smooth curve, we mark the initial point with an "o" and the points of contact with the ramp with an "*".

```
function ch2ex3
% The ball is at (x(t),y(t)).
% Here y(1) = x(t), y(2) = x'(t), y(3) = y(t), y(4) = y'(t).

% Set initial height and coefficient of restitution:
y0 = [0; 0; 2; 0];
k = 0.35;

% Plot the initial configuration:
fill([0 0 1],[0 1 0],[0.8 0.8 0.8]);
axis([-0.1 1.1 0 y0(3)])
hold on
plot(0,y0(3),'ro'); % Mark the initial point.

options = odeset('Events',@events);
% Accumulate the path of the ball in xplot,yplot.
xplot = [];
yplot = [];
tstar = 0;
while 1
    tspan = linspace(tstar,tstar+1,20);
    [t,y,te,ye,ie] = ode23(@odes,tspan,y0,options);
    % Accumulate the path.
    xplot = [xplot; y(:,1)];
    yplot = [yplot; y(:,3)];
    if isempty(ie)              % Extend the interval.
        tstar = t(end);
        y0 = y(end,:);
    elseif ie(end) == 1        % Ball bounced.
        plot(ye(end,1),ye(end,3),'r*'); % Mark the bounce point.
        if (te(end) - tstar) < 0.01*tstar
            fprintf('Bounces accumulated at x = %g.\n',ye(end,1))
            break;
        end
        if abs(ye(end,1) - 1) < 0.01
            break;
```

```
        end
        tstar = te(end);
        y0 = [ye(end,1); -k*ye(end,4); ye(end,3); k*ye(end,2)];
    elseif ie(end) == 2     % Reached end of ramp.
        break;
    end
end
plot(xplot,yplot);

%=======================================================
function dydt = odes(t,y)
dydt = [y(2); 0; y(4); - 9.81];

function [value,isterminal,direction] = events(t,y)
value = [ y(3) - (1 - y(1))
            y(1) - 1         ];
isterminal = [1; 1];
direction  = [-1; 0];
```

■ **EXERCISE 2.23**

Dormand (1996, p. 114) offers an amusing exercise about the ejection of a cork from a bottle containing a fermenting liquid. Let $x(t)$ be the displacement of the cork at time $t$ and
let $L$ be the length of the cork. While $x(t) \le L$ (the cork is still in the neck of the bottle),
the displacement satisfies

$$\frac{d^2x}{dt^2} = g(1+q)\left[\left(1+\frac{x}{d}\right)^{-\gamma} + \frac{Rt}{100} - 1 + \frac{qx}{L(1+q)}\right]$$

Here the physical parameters are $g = 9.81$, $q = 20$, $d = 5$, $\gamma = 1.4$, and $R = 4$. If
$x(0) = 0$ and $x'(0) = 0$ (i.e., the cork starts at rest) and $L = 3.75$, find when $x(t^*) = L$
(i.e., when the cork leaves the neck of the bottle). Also, what is $x'(t^*)$ (the speed of the
cork as it leaves the bottle)? This is a simple event location problem. Integrate the IVP
until the terminal event $x(t^*) - L = 0$ occurs and then print out $t^*$ and $x'(t^*)$. The only
difficulty is that the MATLAB solvers require you to specify an interval of integration. A
simple way to proceed is to guess that the cork will be ejected by, say, $t = 100$ and ask
the code to integrate over $[0, 100]$. You then need to check that the integration was actually terminated by the event. You can do this by testing whether the integration reached
the end of the interval or by testing whether the array `ie` is empty. If there was no event,
you'll need to try again with a longer interval.

■ **EXERCISE 2.24**

In Chapter 1 we discussed the planar motion of a shot fired from a cannon. This motion is governed by the ODEs

$$y' = \tan(\phi)$$

$$v' = -\frac{g\sin(\phi) + \nu v^2}{v\cos(\phi)}$$

$$\phi' = -\frac{g}{v^2}$$

where $y$ is the height of the shot above the level of the cannon, $v$ is the velocity of the shot, and $\phi$ is the angle (in radians) of the trajectory of the shot with the horizontal. The independent variable $x$ measures the distance from the cannon. The constant $\nu$ represents air resistance (friction) and $g = 0.032$ is the appropriately scaled gravitational constant. A natural question is to determine the range of a shot for given muzzle velocity $v(0)$ and initial angle of elevation $\phi(0)$. The initial height $y(0) = 0$, so we want the first $x^* > 0$ for which $y(x)$ vanishes. This is an event location problem for which a terminal event occurs at the initial point. Chapter 1 discusses another natural question that leads to a BVP: For what initial angle $\phi(0)$ does the cannon have a given range? Figure 1.3 shows two trajectories for muzzle velocity $v(0) = 0.5$ that have range 5 when $\nu = 0.02$. In computing this figure it was found that the two trajectories have $\phi(0) \approx 0.3782$ and $\phi(0) \approx 9.7456$, respectively. For each of these initial angles, solve the IVP with the given values of $v(0)$ and $\nu$ and use event location to verify that the range is approximately 5.

■ **EXERCISE 2.25**

Integrate the pendulum equation

$$\theta'' + \sin(\theta) = 0$$

with initial conditions $\theta(0) = 0$ and $\theta'(0) = 1$ on the interval $[0, 20\pi]$. The solution oscillates a number of times in this interval and so, in order to be more confident of computing an accurate solution, set `RelTol` to $10^{-5}$ and `AbsTol` to $10^{-10}$. Locate all the points $t^*$ where $\theta(t^*) = 0$. Plot $\theta(t)$ and the locations of these events. Compute and display the spacing between successive zeros of $\theta(t)$. In MATLAB this can be done conveniently by computing the spacing with `diff(te)`. The solution is periodic, so if it has been computed accurately then this spacing will be nearly constant. For solutions $\theta(t)$ that are "small", the ODE is often approximated by the linear ODE $x'' + x = 0$. The spacing between successive zeros of $x(t)$ is $\pi$. How does this compare to the spacing you found for the zeros of $\theta(t)$? Reduce the size of $\theta(t)$ by reducing the initial slope to $\theta'(0) = 0.1$. Is the spacing between zeros then closer to $\pi$? What if you reduce the initial slope to $\theta'(0) = 0.01$?

■ **EXERCISE 2.26**

When solving IVPs, typically we make up a table of the solution for given values of the independent variable. However, sometimes we want a table for given values of one of the dependent variables. This can be done using event location. An example is provided by Moler & Solomon (1970), who consider how to compute a table of values of the integral

$$t(x) = \int_{x_0}^{x} \frac{1}{\sqrt{f(s)}} \, ds$$

for a smooth function $f(x) > 0$. Computing the integral is equivalent to solving the IVP

$$\frac{dt}{dx} = \frac{1}{\sqrt{f(x)}}, \quad t(x_0) = 0$$

This ODE will be difficult to solve as $x$ approaches a point where $f(x)$ vanishes. To deal with this, Moler and Solomon first note that $x(t)$ satisfies the IVP

$$\frac{dx}{dt} = \sqrt{f(x)}, \quad x(0) = x_0$$

Because of the square root, this equation also presents difficulties as $x(t)$ approaches a point where $f(x(t))$ has a change of sign or (equivalently) where $x'(t)$ vanishes, so they differentiate it to obtain the IVP

$$x'' = 0.5f'(x), \quad x(0) = x_0, \quad x'(0) = \sqrt{f(x_0)}$$

There is no difficulty integrating this equation through a point where $x'(t)$ vanishes, so such a point can be located easily as a terminal event. We want the values $t_i$ where $x(t)$ has given values $x_i$; that is, we want to locate where the functions $x(t) - x_i$ vanish, a collection of nonterminal events. Write a program to make up a table of values of the integral for given $x_i$. You could hard code the values $x_i$ in your event function, but a more flexible approach is to define an array xvalues of the $x_i$ in the main program and pass it as a global variable to your event function. Check out your program by finding values of the integral for $x_0 = 0$ and xvalues = 0.1:0.1:0.9 when $f(x) = x$. In this use ode45 and integrate over $0 \le t \le 2$. Verify analytically that the integral is $t(x) = 2\sqrt{x}$ and use this to compare the values you compute to the true values. The fact that $f(x)$ vanishes at the initial point does not complicate the numerical integration, but it does mean that there is a terminal event at the initial point. Recall that this is a special case for event location. The solvers report such an event but do not terminate the integration. After checking out your program in this way, modify it to make up a table of values of the integral when $f(x) = 1 - x$ and $f(x) = 1 - x^2$.

■ **EXERCISE 2.27**

The program ch2ex3.m plots the path of a ball bouncing down a ramp when the coefficient of restitution $k = 0.35$. Experiment with the effects of changing $k$. In particular, reduce $k$ sufficiently to see an example of bounces clustering. You might, for example, try $k = 0.6$ and $k = 0.3$.

■ **EXERCISE 2.28**

Suppose that, in the configuration of Example 2.3.5, there is a vertical wall located at the end of the ramp (i.e., where $x(t) = 1$). Modify ch2ex3.m so as to compute the path of the ball then. An easy way to show the wall is to make it the right side of the plot by changing the axis command to axis([-0.1 1 0 y0(3)]). Without the wall, the computation of the path is terminated when the ball reaches the end of the ramp as reported by ie(end) == 2. With a wall at the end of the ramp, this event corresponds to hitting the wall. After hitting the wall, continue integrating with initial values

$$(y_1(t^*), \, -ky_2(t^*), \, y_3(t^*), \, y_4(t^*))$$

If the ball is nearly in the corner, say $y_3(t^*) \leq 0.01$, terminate the computation. Find the path of the ball for $k = 0.35$ and $k = 0.7$.

■ **EXERCISE 2.29**

It is natural to use event location in dfs.m to terminate the integration when $y(t)$ reaches the bottom or the top of the plot window, but that is not the only way to do this. The plot routine displays only the portion of $y(t)$ that lies within the plot window, so all we need to do is stop integrating when the solver steps outside the window. This can be done within the output function. Make yourself a copy of dfs.m and give it a different name. Remove the event function from this program and modify the output function so that the run is terminated if either y(end) < wB or y(end) > wT. This is cheaper than event location because it avoids the expense of locating accurately the time $t^*$ where $y(t)$ reaches either the bottom or top of the plot window.

## 2.3.2 ODEs Involving a Mass Matrix

Some solvers accept systems of ODEs written in forms more general than $y' = f(t, y)$. In particular, the MATLAB IVP solvers accept problems of the form

$$M(t, y)y' = f(t, y) \tag{2.37}$$

involving a mass matrix $M(t, y)$. If $M(t, y)$ is singular then this is a system of *differential algebraic equations* (DAEs). They are closely related to ODEs but differ in important ways. We do not pursue the solution of DAEs in this book beyond noting that the MATLAB codes for stiff ODEs can solve DAEs of index 1 arising in this way (Shampine, Reichelt, & Kierzenka 1999). In Section 2.3.3 we discuss an approach to solving PDEs numerically that leads to systems of ODEs with a great many unknowns. Some of the schemes result in equations of the form (2.37). Example 2.3.11 provides more details about solving such problems.

In simplest use, the only difference when using a MATLAB IVP solver for a problem of the form (2.37) instead of a problem in standard form is that you must inform the solver about the presence of $M(t, y)$ by means of the option `Mass`. If the matrix is a constant then you should provide the matrix itself as the value of this option. The codes exploit constant mass matrices, so this is efficient as well as convenient. Suppose, for example, that you choose an explicit Runge–Kutta method. You provide the ODEs in the convenient form (2.37), but the solver actually applies the Runge–Kutta method to the ODEs

$$y' = F(t, y) \equiv M^{-1} f(t, y)$$

which is in standard form. As the code initializes itself, it computes an $LU$ factorization of $M$. Whenever it needs to evaluate $F(t, y)$, it evaluates $f(t, y)$ and solves a linear system for $F$ using the stored factorization of $M$. For the solvers that proceed in this way, allowing a mass matrix is mainly a convenience for you. However, other solvers modify the methods themselves to include a mass matrix. Without going into the details, it should be no surprise that the iteration matrix for the BDFs is changed from $I - h\gamma J$ to $M - h\gamma J$. The iteration matrix must be factored in any case, so the extra cost of the more general form is unimportant. When the mass matrix is not constant, the modifications to the methods are more substantial. In this case, you must provide a (sub)function for evaluating the mass matrix and pass the name to the solver as the value of `Mass`.

To solve problems of the form (2.37), all that you *must* do is provide the mass matrix necessary to define the ODEs. The solver needs to know whether the mass matrix is singular at the initial point – that is, whether the problem is a DAE. The option `MassSingular` has values `yes`, `no`, and the default of `maybe`. The default causes the solver to perform a numerical test for singularity that you can avoid if you know whether you are solving ODEs or DAEs. For large systems it is essential to take advantage of sparsity and inform the solver about how strongly the mass matrix $M(t, y)$ depends on $y$, matters that are discussed in Example 2.3.11.

### EXAMPLE 2.3.6

MATLAB 6 comes with a demonstration code, `batonode`, illustrating the use of a mass matrix. It is an interesting problem that is not described in the documentation, so we explain it here; it is based on Example 4.3A of Wells (1967). A baton is modeled as two particles of masses $m_1$ and $m_2$ that are fastened to opposite ends of a light straight rod of

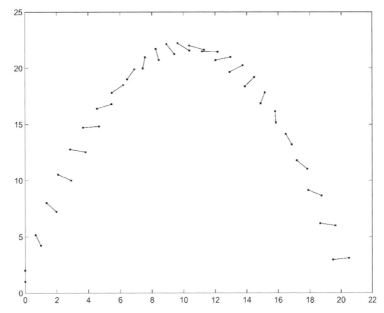

Figure 2.5: Solution of a thrown baton problem with mass matrix $M(t, y)$.

length $L$. The motion of the baton is followed in a vertical plane under the action of gravity. If the coordinates of the first particle at time $t$ are $(X(t), Y(t))$ and if the angle the rod makes with the horizontal is $\theta(t)$, then Lagrange's equations for the motion are naturally expressed in terms of a mass matrix that involves the unknown $\theta(t)$. The ODEs are written in terms of the vector

$$y = (X, X', Y, Y', \theta, \theta')^{\mathrm{T}}$$

and the functions $f(t, y)$ and $M(t, y)$ are defined by the partial listing of `batonode` that follows. (We do not display the comments nor the coding associated with plotting the motion of the baton.) Clearly, it is more convenient to solve this small, nonstiff problem using the natural formulation in terms of a mass matrix than to convert it to the usual standard form. This program exploits the capability of *passing parameter values* through the solver to the functions defining the ODEs and the mass matrix as arguments at the ends of the lists of input variables. This capability is common among the functions of MATLAB, so we content ourselves with an example of its use. The option `MassSingular` could have been set to `no`, but the default test for this is inexpensive with only six equations. There is no advantage to be gained from sparsity when there are only six equations, but to exploit sparsity in this program you need only change `M = zeros(6,6)` to `M = sparse(6,6)`. Figure 2.5 is the result of running the code `batonode` with the title removed from the figure. (Exercise 2.31 has you experiment further with `batonode`.)

```
function batonode
      .
      .
      .
m1 = 0.1;
m2 = 0.1;
L = 1;
g = 9.81;

tspan = linspace(0,4,25);
y0 = [0; 4; 2; 20; -pi/2; 2];

options = odeset('Mass',@mass);
[t y] = ode45(@f,tspan,y0,options,m1,m2,L,g);
      .
      .
      .
% ------------------------------------------

function dydt = f(t,y,m1,m2,L,g)
dydt = [
 y(2)
 m2*L*y(6)^2*cos(y(5))
 y(4)
 m2*L*y(6)^2*sin(y(5))-(m1+m2)*g
 y(6)
 -g*L*cos(y(5))
 ];

% ------------------------------------------

function M = mass(t,y,m1,m2,L,g)
M = zeros(6,6);
M(1,1) = 1;
M(2,2) = m1 + m2;
M(2,6) = -m2*L*sin(y(5));
M(3,3) = 1;
M(4,4) = m1 + m2;
M(4,6) = m2*L*cos(y(5));
M(5,5) = 1;
```

```
M(6,2)  =  -L*sin(y(5));
M(6,4)  =  L*cos(y(5));
M(6,6)  =  L^2;
```

## EXAMPLE 2.3.7

The incompressible Navier–Stokes equations for time-dependent fluid flow in one space
dimension on an interval of length $L$ can be formulated as the system of PDEs

$$\frac{\partial U}{\partial t} + A \frac{\partial U}{\partial z} = C$$

to be solved for $t \geq 0$ and $0 \leq z \leq L$. The vector $U = (\rho \ G \ T)^{\mathrm{T}}$ consists of three unknowns: the density $\rho$, the flow rate $G$, and the temperature $T$. In the system of PDEs,
the vector

$$C = \begin{pmatrix} 0 \\ -KG\left|\dfrac{G}{\rho}\right| - \rho g_a \sin(\theta) \\ \dfrac{a^2 \Phi P_H \kappa}{C_p A_f} \end{pmatrix}$$

and the matrix

$$A = \begin{pmatrix} 0 & 1 & 0 \\ \dfrac{1}{\rho\kappa} - \dfrac{G^2}{\rho^2} & 2\dfrac{G}{\rho} & \dfrac{\beta}{\kappa} \\ -\dfrac{a^2 \beta \bar{T} G}{\rho^2 C_p} & \dfrac{a^2 \beta \bar{T}}{\rho C_p} & \dfrac{G}{\rho} \end{pmatrix}$$

Relevant fluid properties and other problem parameters are described in Thompson &
Tuttle (1986). We use constant values that are found in `ch2ex4.m` with names that are
obvious except possibly `sinth` for $\sin(\theta)$. We also use $\bar{T} = T + 273.15$ and boundary
conditions

$$\rho(0, t) = \rho_0 = 795.5$$
$$T(0, t) = T_0 = 255.0$$
$$G(L, t) = G_0 = 270.9$$

We are interested in computing a steady-state solution of these equations. At steady state
we have $\rho_t = 0$, which implies that $G(z)$ is the constant $G_0$ on the whole interval $0 <
z < L$ and that the PDEs reduce to the ODEs

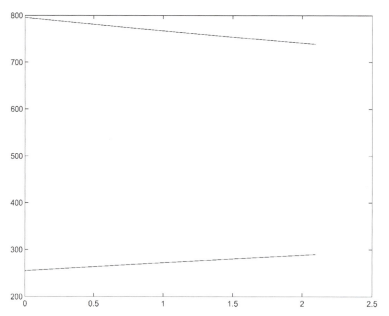

Figure 2.6: Steady-state solution to upper boundary of liquid region.

$$\begin{pmatrix} \dfrac{1}{\rho\kappa} - \dfrac{G_0^2}{\rho^2} & \dfrac{\beta}{\kappa} \\[2ex] -\dfrac{a^2\beta\bar{T}G_0}{\rho^2 C_p} & \dfrac{G_0}{\rho} \end{pmatrix} \begin{pmatrix} \dfrac{d\rho}{dz} \\[2ex] \dfrac{dT}{dz} \end{pmatrix} = \begin{pmatrix} -KG_0 \left| \dfrac{G_0}{\rho} \right| - \rho g_a \sin(\theta) \\[2ex] \dfrac{a^2\Phi P_H \kappa}{C_p A_f} \end{pmatrix}$$

Converting the system of PDEs to a system of ODEs in this way is sometimes referred to as a *continuous space, discrete time* (CSDT) solution. Such a solution may result in an IVP for a system of ODEs or a BVP, depending on the boundary conditions for the PDEs. These equations are often used to model the subcooled liquid portion of a three-phase steam generator in power systems. (Similar but more complicated equations apply to the saturated liquid–steam and pure steam regions.) In such a model the (moving) boundaries between the regions are determined using properties of the equation of state. For example, the ODEs may be integrated from $z = 0$ in the positive $z$ direction until the density $\rho$ is equal to the "liquid-side" saturation density $\rho_{\text{sat}}(T)$. As discussed in Section 2.3.1, we can find where this happens by using an event function

$$g(z, \rho, T) = \rho(z) - \rho_{\text{sat}}(T(z))$$

Strictly for the purposes of illustration and convenience, we use a mock-up of the equation of state:

$$\rho_{\text{sat}}(T) = -3.3(T - 290) + 738$$

Even with the gross simplifications of this model, formulating the IVP is somewhat complicated. However, once we have the IVP, it is easy enough to solve with one of the MATLAB IVP solvers because they accept problems formulated in terms of a mass matrix and they locate events. Invoking `ch2ex4.m` results in output to the screen of

```
Upper boundary at z = 2.09614.
```

and a figure shown here as Figure 2.6. Exercise 2.32 modifies this program in a standard application of the steady-state CSDT model to "pump coast down".

```
function ch2ex4
% Define the physical constants:
global kappa beta a Cp K ga sinth Phi Ph Af G0
kappa = 0.171446272015689e-8;
beta  = 0.213024626664637e-2;
a     = 0.108595374561510e+4;
Cp    = 0.496941623289027e+4;
K     = 10;
ga    = 9.80665;
sinth = 1;
Phi   = 1.1e+5;
Ph    = 797.318;
Af    = 3.82760;
G0    = 270.9;

options = odeset('Mass',@mass,'MassSingular','no','Events',@events);
[z,y,ze,ye,ie] = ode45(@odes,[0 5],[795.5; 255.0],options);
if ~isempty(ie)
     fprintf('Upper boundary at z = %g.\n',ze(end));
end
plot(z,y);

%=================================================================
function dydz = odes(z,y)
global kappa beta a Cp K ga sinth Phi Ph Af G0
rho = y(1);
T = y(2);
dydz = [ (-K*G0*abs(G0/rho) - rho*ga*sinth)
         (a^2 *Phi*Ph*kappa)/(Cp*Af)         ];

function A = mass(z,y)
global kappa beta a Cp K ga sinth Phi Ph Af G0
```

```
rho = y(1);
T = y(2);
A = zeros(2);
A(1,1) = 1/(rho*kappa) - (G0/rho)^2;
A(1,2) = beta/kappa;
A(2,1) = -(a^2 *beta*(T + 273.15)*G0)/(Cp*rho^2);
A(2,2) = G0/rho;

function [value,isterminal,direction] = events(z,y)
isterminal = 1;
direction   = 0;
rho     = y(1);
T       = y(2);
rhosat  = -3.3*(T - 290.0) + 738.0;
value = rho - rhosat;
```

■ **EXERCISE 2.30**

The ODEs of the ozone model presented in Exercise 2.19 appear in a form appropriate for singular perturbation methods. This form involves a mass matrix $M = \text{diag}\{1, \varepsilon\}$. Do the exercise now with the ODEs formulated in terms of a mass matrix. In the computations, take advantage of $M$ being constant.

■ **EXERCISE 2.31**

In the MATLAB demonstration program `batonode`, the length of the baton is 1 and the masses are both 0.1. Make yourself a copy of `batonode` and modify it to compute the motion when the length L is increased to 2 and one mass, say m2, is increased to 0.5. In this you will want to delete, or at least change, the setting of `axis`. What qualitative differences do you observe in the motion?

■ **EXERCISE 2.32**

A standard application of the steady-state CSDT model of Example 2.3.7 is to pump coast down: the pumps are shut down and the inlet flow rate decreases exponentially fast. Solve the CSDT problem for `tD = linspace(0,1,11)` when the inlet flow rate at time `tD(i)` is

$$G_0 = 270.9(0.8 + 0.2e^{-\text{tD}(i)})$$

You can do this by modifying the program `ch2ex4.m` so that it solves for the upper boundary `zU(i)` of the liquid region with this value of $G_0$ in a loop. The flow rate is already a global variable in `ch2ex4.m`, so you can define it in the loop just before calling `ode45`. Use `plot(tD,zU)` to plot the location of the upper boundary as a function of time.

■ **EXERCISE 2.33**

A double pendulum consists of a pendulum attached to a pendulum. Let $\theta_1(t)$ be the angle that the top pendulum makes with the vertical and let $\theta_2(t)$ be the angle the lower pendulum makes with the vertical. Correspondingly, let $m_i$ be the masses of the bobs and $L_i$ the lengths of the pendulum arms. When the only force acting on the double pendulum is gravity with constant acceleration $g$, Borrelli & Coleman (1999) develop the equations of motion

$$(m_1 + m_2)L_1\theta_1'' + m_2L_2\cos(\theta_2 - \theta_1)\theta_2''$$
$$- m_2L_2(\theta_2')^2\sin(\theta_2 - \theta_1) + (m_1 + m_2)g\sin(\theta_1) = 0$$
$$m_2L_2\theta_2'' + m_2L_1\cos(\theta_2 - \theta_1)\theta_1'' + m_2L_1(\theta_1')^2\sin(\theta_2 - \theta_1) + m_2g\sin(\theta_2) = 0$$

In an interesting exercise, Giordano & Weir (1991) model a Chinook helicopter delivering supplies on two pallets slung beneath the helicopter as a double pendulum. They take the mass of the upper pallet to be $m_1 = 937.5$ slugs and the mass of the lower to be $m_2 = 312.5$ slugs. The lengths of the cables are $L_1 = L_2 = 16$ feet and $g = 32$ feet/second$^2$. The hook holding the lower pallet cable is bent open and, if the lower pallet oscillates through an arc of $\pi/3$ radians or more to the open end of the hook, the cable will come off the hook and the pallet will be lost. As a result of a sudden maneuver, the pallets begin moving subject to the initial conditions

$$\theta_1(0) = -0.5, \quad \theta_1'(0) = -1, \quad \theta_2(0) = 1, \quad \theta_2'(0) = 2$$

The differential equations are naturally formulated in terms of a mass matrix, so solve them in this form with `ode45` and default tolerances. Integrate from $t = 0$ to $t = 2\pi$, but terminate the run if the event function $\theta_2(t) - \pi/3$ vanishes (i.e., if the lower pallet is lost) and report when this happens.

Giordano and Weir linearize the differential equations as

$$(m_1 + m_2)L_1\theta_1'' + m_2L_2\theta_2'' + (m_1 + m_2)g\theta_1 = 0$$
$$m_2L_2\theta_2'' + m_2L_1\theta_1'' + m_2g\theta_2 = 0$$

because it is not hard to solve this linear model analytically. The given values of the parameters satisfy $g = 2L_1 = 2L_2$ and $m_1 = 3m_2$. Because of this, the solution has the simple form

$$\theta_1(t) = -\tfrac{1}{2}\cos(2t) - \tfrac{1}{2}\sin(2t)$$
$$\theta_2(t) = \cos(2t) + \sin(2t)$$

An analytical solution provides insight, but it is just as easy to solve the nonlinear model numerically as the linear model. Solve the linear model numerically and compare your numerical solution to this analytical solution. Compare your numerical solution – in particular, the time at which the lower pallet is lost – to the results you obtain with the nonlinear model. You should find that the results of the linear model are quite close to those of the nonlinear model.

### 2.3.3  Large Systems and the Method of Lines

Now we take up issues that are important when solving an IVP for a large system of equations. The MATLAB PSE is not appropriate for the very large systems solved routinely in some areas of general scientific computing, but it is possible to solve conveniently systems that are quite large. The *method of lines* (MOL) is a way of approximating PDEs by ODEs. It leads naturally to large systems of equations that may be stiff. The MOL is a popular tool for solving PDEs, but we discuss it here mainly as a source of large systems of ODEs. MATLAB itself has an MOL code called `pdepe` that solves small systems of parabolic and elliptic PDEs in one space variable and time. Also, the MATLAB PDE Toolbox uses the MOL to solve PDEs in two space variables and time.

The idea of the MOL is to discretize all but one of the variables of a PDE so as to obtain a system of ODEs. This approach is called a *semidiscretization*. Typically the spatial variables are discretized and time is used as the continuous variable. There are several popular ways of discretizing the spatial variable. Two will be illustrated with an example taken from Trefethen (2000).

### EXAMPLE 2.3.8

The one-way wave (also known as the advection or convection) PDE

$$u_t + c(x)u_x = 0, \quad c(x) = \tfrac{1}{5} + \sin^2(x - 1)$$

is solved for $0 \leq x \leq 2\pi$ and $0 \leq t \leq 8$. Trefethen (2000) considers initial values $u(x, 0) = e^{-100(x-1)^2}$ and periodic boundary conditions $u(0, t) = u(2\pi, t)$. As he notes, the initial profile $u(x, 0)$ is not periodic, but it decays so fast near the ends of the interval that it can be regarded as periodic. Also, solutions of this PDE are waves that move to the right. As seen in Figure 2.7, the peak that is initially at $x = 1$ does not reach the boundary of the region in this time interval. For later use we note that a boundary condition must be specified at the left end of the interval, but none is needed at the right. In the MOL, a grid $x_1 < x_2 < \cdots < x_N$ is chosen in the interval $[0, 2\pi]$ and functions $v_m(t)$ are used to approximate $u(x_m, t)$ for $m = 1, 2, \ldots, N$. These functions are determined by a system of ODEs

$$\frac{dv_m}{dt} = -c(x_m)(Dv)_m$$

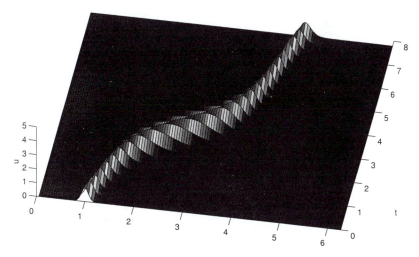

Figure 2.7: Solution of $u_t + c(x)u_x = 0$ with the MOL.

with initial values

$$v_m(0) = u(x_m, 0)$$

for $m = 1, 2, \ldots, N$. If $(Dv)_m \approx u_x(x_m, t)$, then this is an obvious approximation of the PDE. Trefethen's delightful little book is about spectral methods. In that approach, the partial derivative is approximated at $(x_m, t)$ by interpolating the function values $v_1(t), v_2(t), \ldots, v_N(t)$ with a trigonometric polynomial in $x$, differentiating this polynomial with respect to $x$, and evaluating this derivative at $x_m$. It is characteristic of a spectral method that derivatives are approximated using data from the whole interval. When the mesh points are equally spaced, this kind of approximation is made practical using the fast Fourier transform (FFT). Just how this is done is not important here and the details can be found in the discussion of Trefethen's program p6. Our program ch2ex5.m is much like p6 except that Trefethen uses a constant-step method with memory to integrate the ODEs and we simply use ode23. The general-purpose IVP solver is a little more expensive to use, but it avoids the issue of starting a method with memory, adapts the time steps to the solution, and controls the error of the time integration. Moreover, someone else has gone to all the trouble of writing a quality program to do the time integration!

```
function ch2ex5
global ixindices mc

% Spatial grid, initial values:
N = 128;
```

```
x = (2*pi/N)*(1:N);
v0 = exp(-100*(x - 1) .^ 2);

% Quantities passed as global variables to f(t,v):
ixindices = i * [0:N/2-1 0 -N/2+1:-1]';
mc = - (0.2 + sin(x - 1) .^ 2)';

% Integrate and plot:
t = 0:0.30:8;
[t,v] = ode23(@f,t,v0);
mesh(x,t,v), view(10,70), axis([0 2*pi 0 8 0 5])
ylabel t, zlabel u, grid off


%===========================================================
function dvdt = f(t,v)
global ixindices mc
dvdt = mc .* real(ifft(ixindices .* fft(v)));
```

When using a constant step size as Trefethen does, a key issue is selecting a time step small enough to compute an accurate solution. When using a solver like ode23, the issue is choosing output points that result in an acceptable plot because the solver selects time steps that provide an accurate solution and returns answers wherever we like. It would be natural to use the default output at every step, but storage can be an issue for a large system. You can reduce the storage required by asking for answers at specific points. Alternatively, you can gain complete control over the amount of output by writing an output function as illustrated in Example 2.3.3. In ch2ex5.m we specify half as many output points as Trefethen because this is sufficient to provide an acceptable plot and it significantly reduces the size of the output file. For just this reason the PDE solver pdepe requires you to specify not only the fixed mesh in the spatial variable but also the times at which you want output. In a situation like this, you should not ask for output in the form of a solution structure because such structures contain the solution at every step as well as considerably more information that is needed for evaluating the continuous extension.

It is commonly thought that any system of ODEs with a moderate to large number of unknowns that arises from MOL is stiff, but the matter is not that simple. The system of 128 ODEs arising in the spectral discretization of this wave equation is at most mildly stiff. This is verified by using the stiff solver ode15s instead of the nonstiff solver ode23; the change *increases* the run time by more than a factor of 5. When no substantial advantage is to be gained from the superior stability of its formulas, ode15s is often (much) more expensive than one of the solvers intended for nonstiff IVPs. By default, ode15s approximates the Jacobian matrix numerically. This is expensive because, for a system of $N$

equations, it makes $N$ evaluations of the ODE function $f(t, v)$ for this purpose (and a few more if it has doubts about the quality of the approximation). We remark that ode15s handles this much more efficiently than most stiff solvers because it saves Jacobians and forms new ones only when it believes this is necessary for the efficient evaluation of the implicit formulas. When solving nonstiff problems, the step size is generally small enough that only a few Jacobians are formed. To explore this, we set the option Stats to on and ran the modified ch2ex5.m to obtain

```
>> ch2ex5
199 successful steps
6 failed attempts
328 function evaluations
1 partial derivatives
34 LU decompositions
324 solutions of linear systems
```

(*Note:* "*LU* decompositions" is synonymous with "*LU* factorizations".) The implicit formulas are evaluated with a remarkably small number of function evaluations per step. The solver made only one Jacobian (partial derivative) evaluation and reused it each of the 34 times it formed and factored an iteration matrix. By these measures, ode15s solves the nonstiff IVP quite efficiently. Indeed, ode23 took 315 steps and made 946 function evaluations. However, ode15s is slower because it had to solve many systems involving 128 linear equations. They are solved very efficiently in MATLAB, but the extra overhead of the BDF code makes it less efficient than the Runge–Kutta code for this problem. On the other hand, the higher-order explicit Runge–Kutta code ode45 is less efficient than ode15s. Because its formulas have smaller stability regions, ode45 is less efficient for this problem than ode23.

## EXAMPLE 2.3.9

The spectral approximation of $u_x$ in Example 2.3.8 is very accurate when the function $u(x, t)$ is smooth. Finite difference approximations are a more common alternative. They are of (much) lower order of accuracy, but they are local in nature and so require (much) less smoothness of the solution. It is worthy of comment that the one-way wave equation can have solutions with discontinuities in the spatial variable, a matter that receives a great deal of attention in the numerical solution of such PDEs. A simple, yet reasonable, finite difference scheme for this equation is a first-order upwind difference approximation to the spatial derivative. Again we use a mesh in $x$ of $N$ points with equal spacing of $h$. Because $c(x) > 0$, the wave is moving to the right and the scheme results in

$$\frac{dv_m}{dt} = -c(x_m)\left(\frac{v_m - v_{m-1}}{h}\right)$$

Trefethen assumed periodic boundary conditions because doing so simplified the use of spectral approximations. At the left boundary this assumption leads to

$$\frac{dv_1}{dt} = -c(x_1)\left(\frac{v_1 - v_N}{h}\right) \qquad (2.38)$$

We'll return to this matter, but right now let us also change the boundary condition to simplify the application of the finite difference scheme. With the given initial data, it is just as plausible to use the boundary condition $u_x(0, t) = 0$. This boundary condition is approximated by $v_1'(t) = 0$. With a first-order approximation to $u_x(x, t)$, we need a fine mesh to obtain a numerical solution comparable to that of Figure 2.7. The program ch2ex6.m approximates this figure by taking $N = 1000$. Notice that with such a fine mesh we do not plot the solution at all the mesh points.

The finite difference program is straightforward, but new issues arise when we solve problems involving a large number of equations (here 1000). The problems resulting from MOL with a fine space discretization can be stiff. Indeed, the parabolic PDEs solved with pdepe are stiff when the spatial mesh is at all fine. The IVP of this example is not stiff, but it is simple enough that it shows clearly how to proceed when we treat it as a stiff problem. If we solve it with a code intended for stiff problems, the code must form Jacobians. With $N$ equations, this matrix is $N \times N$. Thousands of equations are common in this context, making it important to account for the sparsity of the equations with respect to both storage and the efficiency of solving linear systems. Sparse matrix technology is fully integrated into the MATLAB PSE, but it is not so readily available in general scientific computing. In fact, the only provision for sparsity in a typical stiff solver for general scientific computing is for *banded Jacobians*. (A banded matrix is one that has all its nonzero entries confined to a band of diagonals about the principal diagonal.) Somehow we must inform the solver of the zeros in the Jacobian $J = \left(\frac{\partial f_m}{\partial v_k}\right)$. The simplest way is to provide the Jacobian matrix analytically as a sparse matrix. This is done much as with mass matrices by means of an option called Jacobian. If the Jacobian is a constant matrix, you should provide it as the value of this option. This could have been done in ch2ex6.m with

```
B = [ [-c_h(2:N); 0] [0; c_h(2:N)] ];
J = spdiags(B,-1:0,N,N);
options = odeset('Jacobian',J);
```

When the Jacobian is not constant, you provide a function that evaluates the Jacobian and returns a sparse matrix as its value. In either case, the solvers recognize that the Jacobian is sparse and deal with it appropriately. When it is not too much trouble to work out an analytical expression for the Jacobian, it is best to do so because approximating Jacobians numerically is a relatively expensive and difficult task. Also, the methods are generally somewhat more efficient when using analytical Jacobians. Still, the default in the MATLAB solvers is to approximate Jacobians numerically because it is so convenient. For large problems it is then very important that you inform the solver of the structure of the Jacobian. The entry $J_{m,k}$ is identically zero when equation $m$ does not depend on component $k$ of the solution. You inform the solver of this by creating a sparse matrix $S$ that has zeros in the places where the Jacobian is identically zero and ones elsewhere. It is provided as the value of the option JPattern. This is done in a very straightforward way in ch2ex6.m. The example shows how you might proceed for more complicated patterns, but for this particular set of ODEs the nonzero entries occur on diagonals, making it natural to define the pattern with spdiags. This could have been done in ch2ex6.m with

```
S = spdiags(ones(N,2),-1:0,N,N);
S(1,1) = 0;
```

You are asked in Exercise 2.34 to verify that this IVP is not stiff. Nonetheless, we solve it with ode15s in ch2ex6.m to show how to solve a large stiff problem. The Jacobian is approximated by finite differences, but the sparsity pattern is supplied to make this efficient. It would be more efficient to provide an analytical expression for the constant Jacobian matrix as the value of the Jacobian option.

```
function ch2ex6
global c_h

% Spatial grid, initial values:
N = 1000;
h = 2*pi/N;
x = h*(1:N);
v0 = exp(-100*(x - 1) .^ 2);
c_h = - (0.2 + sin(x - 1) .^ 2)' / h;

% Sparsity pattern for the Jacobian:
S = sparse(N,N);
for m = 2:N
    S(m,m-1) = 1;
    S(m,m) = 1;
end
options = odeset('JPattern',S);

% Integrate and plot:
t = 0:0.30:8;
[t,v] = ode15s(@f,t,v0,options);
```

```
pltspace = ceil(N/128);
x = x(1:pltspace:end);
v = v(:,1:pltspace:end);
surf(x,t,v), view(10,70), axis([0 2*pi 0 8 0 5])
ylabel t, zlabel u, grid off


%==================================================
function dvdt = f(t,v)
global c_h
dvdt = c_h .* [0; diff(v)];
```

Now we can understand better the role of the boundary conditions. With periodic boundary conditions, this example does not have a banded Jacobian because the first equation depends on $v_N$. Because of this, typical codes for stiff problems in general scientific computing cannot solve this IVP directly. In contrast, with the stiff solvers of MATLAB, all we must do is follow the definition of S in ch2ex6.m with the two statements

```
    S(1,1) = 1;
    S(1,N) = 1;
```

Exercise 2.35 asks you to solve this IVP with periodic boundary conditions.

If no attention is paid to sparsity, forming an approximate Jacobian by finite differences costs (at least) one evaluation of the ODE function $f(t, v)$ for each of the $N$ columns of the matrix. The MATLAB solvers use a development of an algorithm due to Curtis et al. (1974) that can reduce this cost greatly by taking into account entries of the Jacobian that are known to be zero. For instance, if the Jacobian is banded and has $D$ diagonals, this algorithm evaluates the Jacobian with only $D$ evaluations of the ODE function $f(t, v)$, *no matter what the value of N*. Informing the solver of the structure of the Jacobian in ch2ex6.m allows it to approximate the Jacobian in only *two* evaluations of the ODE function $f(t, v)$ instead of the $D = N = 1000$ evaluations required if the structure is ignored. Clearly this is a matter of great importance for efficiency, quite aside from its importance for reducing the storage and cost of solving the linear systems involving the iteration matrix.

### EXAMPLE 2.3.10

As we have just seen, one way to speed up the numerical approximation of Jacobians is to use sparsity to reduce the number of function evaluations. Another is to evaluate the functions faster. Vectorization is a valuable tool for speeding up MATLAB programs, and often it is not much work to vectorize a program. When the solvers approximate a Jacobian, they evaluate $f(t, y)$ for several vectors $y$ at the same value of $t$. This can be exploited by coding the function so that, if it is called with arguments (t,[y1 y2 ... ]), it will compute and return [f(t,y1) f(t,y2) ... ]. You tell the solver that you have done this by setting the option Vectorized to on. Because all the built-in functions of MATLAB are vectorized, vectorizing $f(t, y)$ is often a matter of treating y as an array of column vectors and putting a dot before operators to make them apply to arrays. The brusselator problem is a pair of coupled PDEs solved in Hairer & Wanner (1991, pp. 6–8) by the MOL with finite difference discretization of the spatial variable. It is also solved by one of the demonstration programs that is included in MATLAB. When invoked as brussode(N), this program forms and integrates a system of $2N$ ODEs. The default value of the parameter $N$ is 20. Here we consider just one line of the program to make the point about vectorizing $f(t, y)$. Some of the components of dydt are evaluated in a loop on i that might have been coded as

```
dydt(i) = 1 + y(i+1)*y(i)^2 - 4*y(i) + ...
             c*(y(i-2) - 2*y(i) + y(i+2));
```

This processes a column vector y(:). A vector version must perform the same operations on the columns of an array y(:,:). It is coded in brussode as

```
dydt(i,:) = 1 + y(i+1,:).*y(i,:).^2 - 4*y(i,:) + ...
               c*(y(i-2,:) - 2*y(i,:) + y(i+2,:));
```

The dot before the multiplication and exponentiation operators makes the operations apply by components to arrays. Adding the scalar 1 to an array is interpreted as adding it to each entry of the array. This is easy enough, but vectorization is unimportant for this example. The sparsity structure of the Jacobian is such that only a few function evaluations are needed for each Jacobian and only a few Jacobians are formed. Indeed, by setting the option Stats to on, we found that only two Jacobians were formed for the whole integration. Whether vectorization is helpful depends on the sparsity pattern of the Jacobian and how expensive it is to evaluate the function. A similar vectorization is usually quite helpful when solving BVPs with bvp4c because of its numerical method, so we return to this matter in Chapter 3.

To illustrate a vectorization that is not so straightforward, suppose that we now replace ode23 in ch2ex5.m by ode15s and so wish to vectorize

```
dvdt = mc .* real(ifft(ixindices .* fft(v)));
```

The fast Fourier transform routines are vectorized, so fft(v) works fine when v has more than one column. We get into trouble when we form ixindices .* fft(v) because ixindices is a column vector and the array sizes do not match up. We need to do this array multiplication on each of the columns of fft(v). We could do this in a loop.

In such a loop we must take account of the fact that the solver calls the function with an array $v$ that generally has N columns but sometimes has only one. A more efficient way to evaluate the derivatives is to make `ixindices` a matrix with N columns, all the same. This is done easily with `repmat`:

```
temp1 = i * [0:N/2-1 0 -N/2+1:-1]';
ixindices = repmat(temp1,1,N);
```

To match up the dimensions properly for the array multiplication of `ixindices` and `fft(v)`, we just use `size` to find out how many columns there are in $v$ and use the corresponding number of columns of `ixindices` in the multiplication. The same change must be made with the array `mc`. The evaluation of `dvdt` then becomes

```
nc = size(v,2);
dvdt = mc(:,1:nc) .* real(ifft(ixindices(:,1:nc) .* fft(v)));
```

Vectorizing $f(t,v)$ has little effect for this particular IVP because it is not stiff and `ode15s` forms only one Jacobian. However, the point of this example is to show how complications can arise in vectorization.

## EXAMPLE 2.3.11

Mass matrices arise naturally when a Galerkin method is used for the spatial discretization. To illustrate this and so introduce a discussion of associated issues for large stiff systems, we consider an example of Fletcher (1984). He solves the heat equation,

$$u_t = u_{xx}, \quad 0 \le x \le 1$$

with initial value

$$u(x, 0) = x + \sin(\pi x)$$

and boundary values

$$u(0, t) = 0, \quad u(1, t) = 1$$

An approximate solution is sought in the form

$$v(x, t) = \sum_m S_m(x) v_m(t)$$

For a given value of $t$, this approximation satisfies the PDE with a residual

$$R(x, t) = v_t(x, t) - v_{xx}(x, t)$$

An inner product of two functions continuous on $[0, 1]$ is defined by

$$(f, g) = \int_0^1 f(x)g(x)\, dx$$

The Galerkin method projects $u(x, t)$ into the space of functions of the form $v(x, t)$ by requiring the residual of the approximation to be orthogonal to the basis functions $S_k(x)$; that is,

$$(R, S_k) = (v_t, S_k) - (v_{xx}, S_k) = 0$$

for each $k$. If each shape function $S_m(x)$ is a piecewise-linear function with $S_m(x_m) = 1$ and $S_m(x_j) = 0$ for $j \neq m$, then

$$v(x_m, t) = v_m(t) \approx u(x_m, t)$$

Using the form of $v(x, t)$ and performing an integration by parts leads to the equation

$$\sum_m (S_m, S_k)\frac{dv_m}{dt} + \sum_m \left(\frac{dS_m}{dx}, \frac{dS_k}{dx}\right) v_m = 0$$

With a constant mesh spacing $h$, working out the inner products leads to the ODEs

$$\frac{1}{6}\frac{dv_{m-1}}{dt} + \frac{4}{6}\frac{dv_m}{dt} + \frac{1}{6}\frac{dv_{m+1}}{dt} = \frac{v_{m-1} - 2v_m + v_{m+1}}{h^2}$$

for $m = 1, 2, \ldots, N$. From the boundary conditions we have

$$v_0(t) = 0, \quad v_{N+1}(t) = 1$$

These quantities appear only in the first and last ODEs. Notice that the last ODE is inhomogeneous for this reason. From the initial condition, $v_m(0) = u(x_m, 0)$ for $m = 1, 2, \ldots, N$.

Many problems are naturally formulated in terms of mass matrices. If $M(t, y)$ does not depend strongly on $y$, then it is not difficult to modify the numerical methods to take it into account and the presence of a mass matrix has little effect on the computation. For a large system it is essential to account for the structure of the mass matrix as well as the structure of the Jacobian. Although it is often both convenient and efficient to use the standard form (2.37), the typical general-purpose BDF code in general scientific computing does not provide for mass matrices. Storage and the user interface are important reasons for this. For example, when the mass matrix is constant, the iteration matrix for the BDFs is changed from $I - h\gamma J$ to $M - h\gamma J$. The change complicates specification of the matrices and management of storage because the structures of $M$ and $J$ must be merged in forming the iteration matrix. This complication is not present in the MATLAB IVP solvers because the matrices are provided as general sparse matrices and the PSE deals automatically with the storage issues.

The modifications to the algorithms due to the presence of a mass matrix depend on its form. The weaker the dependence of $M(t, y)$ on $y$, the better. The most favorable case is

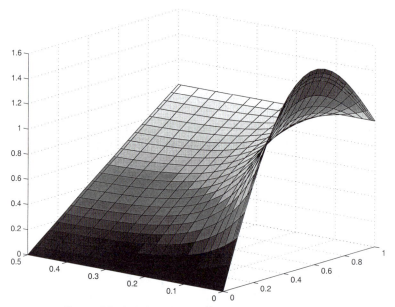

Figure 2.8: Solution of the differential equation $u_t = u_{xx}$.

a constant mass matrix. The solver is told of this by providing the matrix as the value of the Mass option. If the mass matrix is not constant, the option MStateDependence is used to inform the solver about how strongly $M(t, y)$ depends on $y$. This is an issue only for moderate to large systems. There are three values possible for this option. The most favorable is none, indicating a mass matrix that is a function of $t$ only – that is, the matrix has the form $M(t)$. The default value is weak. The most difficult in both theory and practice is strong. If $M(t, y)$ depends strongly on $y$, then the standard form (2.37) is much less advantageous and the computation resembles much more closely what must be done for a general implicit system of the form

$$F(t, y, y') = 0 \tag{2.39}$$

Solving efficiently a large system of ODEs when the mass matrix has a strong state dependence is somewhat complicated and the default of a weak state dependence is usually satisfactory, so we leave this case to the MATLAB documentation.

There are a number of codes in general scientific computing (DASSL is a popular one; see Brenan et al. 1996) that accept problems of the form (2.39), and a future version of MATLAB will include a code for such problems. This general form includes differential algebraic equations (DAEs). They can be much more difficult than ODEs in both theory and practice. An obvious theoretical difficulty is that an initial value $y(t_0)$ may not be sufficient to specify a solution – we must find an initial slope $y'(t_0)$ that is *consistent* in the sense that equation (2.39) is satisfied with arguments $t_0$, $y(t_0)$, and $y'(t_0)$. It is pretty obvious from the form of the equation that the solvers need $\frac{\partial F}{\partial y'}$ as well as $\frac{\partial F}{\partial y}$, complicating the user interface and storage management.

For the present example it is convenient to supply both the mass matrix and the Jacobian matrices directly because they are constant. Indeed, it is convenient to use the Jacobian matrix in evaluating the ODEs. Figure 2.8 shows the output of ch2ex7.m, but a better understanding of the solution is obtained by using the Rotate 3D tool to examine the surface from different perspectives.

```
function ch2ex7
global J h

N = 20;
h = 1/(N+1);
x = h*(1:N);
v0 = x + sin(pi*x);
e = ones(N,1);
M = spdiags([e  4*e e],-1:1,N,N)/6;
J = spdiags([e -2*e e],-1:1,N,N)/h^2;
options = odeset('Mass',M,'Jacobian',J);
[t,v] = ode15s(@f,[0 0.5],v0,options);
% Add the boundary values:
x = [0 x 1];
npts = length(t);
v = [zeros(npts,1) v ones(npts,1)];
surf(x,t,v)

%===========================================
function dvdt = f(t,v)
global J h
dvdt = J*v;
dvdt(end) = dvdt(end) + 1/h^2;
```

■ **EXERCISE 2.34**

Modify ch2ex6.m to use ode23. Using tic and toc, measure how much the run time is affected by this change, hence how stiff the IVP appears to be. It can't be very stiff if you can solve it at all with ode23.

■ **EXERCISE 2.35**

Example 2.3.9 points out that it is easy to deal with periodic boundary conditions because ode15s provides for general sparse matrices. See for yourself by modifying ch2ex6.m so as to solve the problem with periodic boundary conditions. The example explains how to form the sparsity pattern $S$ in this case. Only the first component in `dvdt` must be altered so that it corresponds to the periodic boundary condition (2.38).

### ■ EXERCISE 2.36

The "quench front" problem of Laquer & Wendroff (1981) models a cooled liquid rising on a hot metal rod by the PDE

$$u_t = u_{xx} + g(u)$$

for $0 \le x \le 1$, $0 < t$. Here

$$g(u) = \begin{cases} -Au & \text{if } u \le u_c \\ 0 & \text{if } u_c < u \end{cases}$$

with $A = 2 \cdot 10^5$ and $u_c = 0.5$. The boundary conditions are $u(0, t) = 0$ and $u_x(1, t) = 0$. The initial condition is

$$u(x, 0) = \begin{cases} 0 & \text{if } 0 \le x \le 0.1 \\ \frac{x - 0.1}{0.15} & \text{if } 0.1 < x < 0.25 \\ 1 & \text{if } 0.25 \le x \le 1 \end{cases}$$

We'll approximate the solution of this PDE by the MOL with finite differences. The computations will help us understand stiffness.

For an integer $N$, let $h = 1/N$ and define the grid $x_m = mh$ for $m = 0, 1, \ldots, N + 1$. We approximate the PDE by letting $v_m(t) \approx u(x_m, t)$ and discretizing in space with central differences to obtain

$$\frac{dv_m}{dt} = \frac{v_{m+1} - 2v_m + v_{m-1}}{h^2} + g(v_m)$$

As written, the equation for $m = 1$ requires $v_0(t)$. To satisfy the boundary condition $u(0, t) = 0$, we define $v_0(t) = 0$ and eliminate it from this equation. Similarly, the equation for $m = N$ requires $v_{N+1}(t)$. We approximate the boundary condition at $x = 1$ by central differences as

$$0 = u_x(1, t) \approx \frac{v_{N+1} - v_{N-1}}{2h}$$

Accordingly, we take $v_{N+1} = v_{N-1}$ and eliminate it in the equation for $m = N$. Initial values for the ODEs are provided by $v_m(0) = u(x_m, 0)$. Solve this IVP for the equations $m = 1, 2, \ldots, N$ with $N = 50$. In solving this problem, take advantage of the fact that the Jacobian $J$ is a constant, sparse matrix. You can do this as in `ch2ex7.m`, though here $J_{N, N-1} = 2h^{-2}$. It is convenient and efficient to use $J$ when evaluating the ODEs. You might also vectorize the evaluation of the $g(v_m)$. For this you might find it helpful to ask yourself what the result is of `any(w <= 0,2)` when w is a column vector. Control the output by taking `tspan = 0:0.001:0.006`. If you then plot the computed solution v against the mesh x, you will obtain solution profiles that show how the solution evolves in time. Solve the IVP using each of ode15s and ode23. Use tic and toc to measure the costs of integration. You will find that this IVP is not stiff – ode23 is much more efficient than ode15s. People familiar with the MOL are likely to find this surprising. To see what they might have expected, drop the inhomogeneous term $g(u)$ from the ODEs and solve with tspan = 0:0.1:0.6. You will find that this new IVP is (moderately) stiff.

What's going on? The inhomogeneous term $g(u)$ is not smooth. This results in solution components $v_m(t)$ that are not smooth as they pass through the value $u_c$. For this problem, that happens only once for any $m$. The solver must locate this point and use a small step size to pass it, but an isolated point where a solution is not smooth has little impact on an integration. The snag here is that this happens at different times for the different components, and there are a lot of components. Remember, for an IVP to be stiff, the solution must be smooth so that the step size of an explicit method is limited by stability. Stability does not limit the step size significantly for the quench problem, so ode15s is much less efficient than ode23 even though we have a constant, analytical Jacobian.

## 2.3.4  Singularities

Standard codes for IVPs expect some smoothness, and when it is not present at an isolated point we must supplement the codes with some analytical work. The idea is to approximate the solution of interest near the singular point with a series or asymptotic expansion and approximate it elsewhere with a standard IVP solver. The expression "solution of interest" alludes to the possibility of more than one solution at a singular point. Singularities are much more common for BVPs, in part because problems are often set on an infinite interval, so we show here how to proceed with a single example and a couple of exercises and later return to the matter in Chapter 3, where many examples are discussed at length.

### EXAMPLE 2.3.12

A classical analysis (Lighthill 1986, pp. 103ff) of the collapse of a spherical cavity in a liquid leads to the IVP

$$(y')^2 = \tfrac{2}{3}(y^{-3} - 1), \quad y(0) = 1 \tag{2.40}$$

The nondimensional variables here are the time $x$ and the radius of the cavity $y$. The integration is to terminate when $y$ vanishes, representing total collapse of the cavity. In standard form the ODE is

$$\frac{dy}{dx} = -\sqrt{\tfrac{2}{3}(y^{-3} - 1)}$$

where the minus sign is taken because the equation is to model *collapse* of the cavity. This IVP presents two kinds of difficulties. The existence and uniqueness result for IVPs supposes that the ODE function $f$ in $y' = f(x, y)$ is smooth in a region containing the initial point $(0, 1)$. That is not the case here because the initial point is on the boundary of the region where $f$ is defined. This leaves open the possibility of more than one solution and, in fact, there is an obvious solution $y(x) \equiv 1$ in addition to the decreasing solution that we expect physically. Certainly we must consider how to compute the "right" solution. The other difficulty is that, at the time $x_c$ of total collapse, we have $y(x_c) = 0$ and we find from the ODE that $y'(x_c) = -\infty$.

We approximate the solution $y(x)$ for $0 \le x \le d$ with a Taylor series. We then use $y_d \approx y(d)$ computed in this way to start a numerical integration for $x \ge d$ where the ODE is not singular. The symbolic capabilities of MATLAB make it easy to obtain the series. Taking into account the initial value, we look for an approximation of the form $y(x) = 1 + ax + bx^2 + \cdots$. The script

```
syms y x a b res
y = 1 + a*x + b*x^2;
res = taylor(diff(y)^2 - (2/3)*(1/y^3 - 1),3)
```

substitutes this form into the ODE and computes the first three terms of a Taylor expansion of the residual. The result of this computation is

```
res = a^2+(4*a*b+2*a)*x+(4*b^2+2*b-4*a^2)*x^2
```

To satisfy the ODE as well as possible, we must choose the coefficients so as to make as many terms as possible vanish, starting from the lowest order. To remove the constant term in the residual, we must take $a = 0$, which also removes the term in $x$. To remove the term in $x^2$, we have two choices for $b$, namely $b = 0$ and $b = -0.5$. Accordingly, this singular IVP appears to have exactly two solutions that can be expanded in Taylor series. One choice for $b$ corresponds to the solution $y(x) \equiv 1$ that we have already recognized. The other solution,

$$y(x) = 1 - \tfrac{1}{2}x^2 + \cdots$$

is decreasing for small $x$. It appears then that there is a unique solution with the expected physical behavior. Continuing in this way, it is found easily that

$$y(x) = 1 - \frac{1}{2}x^2 - \frac{1}{6}x^4 - \frac{19}{180}x^6 + \cdots$$

In ch2ex8.m we approximate $y(x)$ by the first three terms of this series and then estimate the error of the approximation by the next term. In this way we estimate that the relative error of this approximation to $y(d)$ is about $10^{-7}$ when $d = 0.1$.

The difficulty with calculating the vertical slope at the end of the integration is resolved by interchanging the independent and dependent variables. This is a technique familiar in the theory of ODEs that can be quite useful numerically. We must be careful that interchanging variables is permissible. We can use $y$ as the independent variable if we stay away from the initial point because the ODE shows that $y(x)$ is strictly decreasing. We cannot use $y$ as the independent variable there, because $y'(x)$ vanishes at $x = 0$ and we'd have the same problem of a vertical slope after interchanging variables. For this reason we integrate

$$\frac{dx}{dy} = -\sqrt{\frac{3y^3}{2(1 - y^3)}}$$

with initial value $x = d$ from $y = y_d$ to $y = 0$. The time of total collapse is the value of $x$ at $y = 0$. After this analytical preparation of the problem, it is easily solved as in ch2ex8.m. To plot the solution over the whole interval, we must augment the numerical solution with values obtained by evaluating the series on the interval $[0, d]$. For this particular problem, a smooth graph is obtained by adding only one value, the initial value $y(0) = 1$. You might wish to see what happens if you make the initial point $d$ much larger or much smaller than $d = 0.1$.

```
function ch2ex8
d = 0.1;
yd = 1 - (1/2)*d^2 - (1/6)*d^4;
erryd = (19/180)*d^6;
fprintf('At d = %g, the error in y(d) is about %5.1e.\n',d,erryd);

[y,x] = ode45(@ode,[yd 0],d);
fprintf('Total collapse occurs at x = %g.\n',x(end));

% Augment the arrays with y = 1 at x = 0 for the plot
% and plot with the original independent variable x.
y = [1; y];
x = [0; x];
plot(x,y)

%=============================================================
function dxdy = ode(y,x)
dxdy = - sqrt(3*y^3/(2*(1 - y^3)));
```

In addition to Figure 2.9, this program results in the output

```
>> ch2ex8
At d = 0.1, the error in y(d) is about 1.1e-007.
Total collapse occurs at x = 0.914704
```

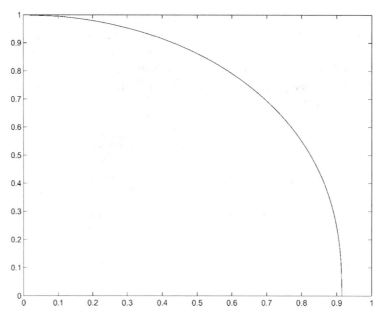

Figure 2.9: Solution of the cavity collapse problem.

### ■ EXERCISE 2.37

Davis (1962, pp. 371ff) discusses the use of Emden's equation

$$\frac{d^2 y}{dx^2} + \frac{2}{x}\frac{dy}{dx} + y^n = 0$$

for modeling the thermal behavior of a spherical cloud of gas. Here $n$ is a physical parameter. The singular coefficient arises when a PDE modeling the thermal behavior is reduced to an ODE by exploiting the spherical symmetry. Symmetry implies that the initial value $y'(0) = 0$ and, in these nondimensional variables, $y(0) = 1$. We expect a solution that is bounded and smooth at the origin, so we expect that we can expand the solution in a Taylor series. Derive the series solution reported by Davis,

$$y(x) = 1 - \frac{x^2}{3!} + n\frac{x^4}{5!} + (5n - 8n^2)\frac{x^6}{3 \cdot 7!} + \cdots$$

When the cloud of gas is a star, the first zero of the solution represents the radius of the star. Davis uses $n = 3$ in modeling the bright component of Capella and reports the first zero to be at about $x = 6.9$. You are to confirm this result. Use the first three terms of the series to compute an approximation to $y(0.1)$ and then use the derivative of these terms to approximate $y'(0.1)$. Estimate the error of each of these approximations by the magnitude of the first term neglected in the series. Using the analytical approximations to $y(0.1)$

and $y'(0.1)$ as the initial values, integrate the ODE with `ode45` until the terminal event $y(x) = 0$ occurs. You'll need to guess how far to go, so be sure to check that the integration terminates because of an event. The approximate solution at $x = 0.1$ is so accurate that it is reasonable to integrate with tolerances more stringent than the default values and thus be more confident in the location of the first zero of $y(x)$. You might, for example, use a relative error tolerance of $10^{-8}$ and an absolute error tolerance of $10^{-10}$.

■ **EXERCISE 2.38**

Kamke (1971, p. 598) states that the IVP

$$y(y'')^2 = e^{2x}, \quad y(0) = 0, \quad y'(0) = 0$$

describes space charge current in a cylindrical capacitor. He suggests approximating the solution with a series of the form

$$y(x) = x^p(a + bx + \cdots)$$

Work out $p$ and $a$. Argue that the IVP has only one solution of this form. Because it is singular at the origin, we must supplement numerical integration of the ODE with this analytical approximation for small $x$. For this you are given that

$$b = \frac{27}{40}a^{-2}$$

If you have not done Exercise 1.7, which asks you to formulate the ODE as a pair of explicit first-order systems, you should do it now. In the present exercise you are to solve numerically the system for which $y'$ is increasing near $x = 0$. Evaluate the series and its derivative to obtain approximations to $y(x_0)$ and $y'(x_0)$ for $x_0 = 0.001$. Integrate the first-order system over $[x_0, 0.1]$ with default error tolerances. Plot the solution that you compute along with an approximate solution obtained from the series.

# Chapter 3

# Boundary Value Problems

## 3.1 Introduction

By itself, a system of ordinary differential equations has many solutions. Commonly a solution of interest is determined by specifying the values of all its components at a single point $x = a$. This point and a direction of integration define an initial value problem. In many applications the solution of interest is determined in a more complicated way. A boundary value problem specifies values or equations for solution components at more than one point in the range of the independent variable $x$. Generally IVPs have a unique solution, but this is not true of BVPs. Like a system of linear algebraic equations, a BVP may not have a solution at all, or may have a unique solution, or may have more than one solution. Because there might be more than one solution, BVP solvers require an estimate (guess) for the solution of interest. Often there are parameters that must be determined in order for the BVP to have a solution. Associated with a solution there might be just one set of parameters, a finite number of possible sets, or an infinite number of possible sets. As with the solution itself, BVP solvers require an estimate for the set of parameters of interest. Examples of the possibilities were given in Chapter 1, and in this chapter others are used to penetrate further into the matter.

The general form of the two-point BVPs that we study in this chapter is a system of ODEs

$$y' = f(x, y, p) \tag{3.1}$$

and a set of boundary conditions

$$0 = g(y(a), y(b), p) \tag{3.2}$$

Here $p$ is a vector of unknown parameters. Parameters may arise naturally in the physical problem that is being modeled or they may be introduced as a part of the process of solving a BVP. Singularities in coefficients of the BVP and problems posed on infinite intervals are not unusual. One way to deal with them involves introducing unknown parameters. The MATLAB BVP solver bvp4c accepts problems with unknown parameters, but many solvers do not, so we explain how to reformulate such BVPs so that there are no unknown parameters. For brevity, we generally do not show $p$ when writing (3.1) or (3.2).

In our study of IVPs we learned how to reformulate ODEs as a system of first-order equations (3.1). It was noted that some IVP solvers accept equations of the form $y'' = f(x, y)$ because they are common in some contexts and there are methods that take advantage of the special form. Similarly, there are popular BVP solvers that work directly with equations of order higher than 1. This can be advantageous, but it complicates the user interface. That is one reason why bvp4c requires first-order systems. Another is that all the IVP solvers of MATLAB require first-order systems.

When there are no unknown parameters, the boundary conditions are said to be *separated* if each of the equations of (3.2) involves just one of the boundary points. If one of the boundary conditions involves solution values at both of the boundary points, then the boundary conditions are said to be *nonseparated*. The solver bvp4c accepts problems with nonseparated boundary conditions, but many solvers do not, so we explain how to reformulate BVPs to separate the boundary conditions. Multipoint BVPs have boundary conditions that are applied at more than two points. Some solvers accept multipoint BVPs, but many (including bvp4c) do not, so we explain how to reformulate such BVPs as two-point BVPs.

It is not always clear what kinds of boundary conditions are appropriate and where they can be applied. We discuss the issues, especially those related to infinite ranges. There is considerable art to solving a BVP with a singularity at an end point or a BVP specified on an infinite range. We look at some simple examples to understand better the possibilities and solve some relatively difficult problems in detail to show how it might be done.

BVPs are much harder to solve than IVPs and any solver might fail, even when provided with a good estimate for the solution and any unknown parameters. Indeed, a BVP solver might also "succeed" when there is no solution! In Section 3.4 we discuss briefly the numerical methods used by popular codes for BVPs. Although the MATLAB BVP solver, bvp4c, is effective, no one method is best for all problems. In particular, the moderate order of its method and the aims of the MATLAB PSE make it inappropriate for problems requiring stringent accuracies or for problems with solutions that have very sharp changes.

After discussing numerical methods for BVPs, we consider examples of solving BVPs with bvp4c. These examples are also used to develop the theory. For instance, they show how to prepare problems for solution with other solvers. They also show how to deal with singularities at an end point and with problems set on an infinite range. The exercises are used similarly, so they should be read even if not actually solved.

# 3.2 Boundary Value Problems

Some examples given in Chapter 1 show that the "facts of life" for BVPs are quite different than for IVPs. In this section we use other examples to learn more about this and to illustrate an approach to analyzing BVPs by means of IVPs.

If the function $f(x, y, y')$ is sufficiently smooth, then the IVP consisting of the equation

$$y'' = f(x, y, y')$$

and initial conditions

$$y(a) = A, \quad y'(a) = s$$

has a unique solution $y(x)$ for $x \geq a$. Two-point BVPs are exemplified by the linear equation

$$y'' + y = 0 \tag{3.3}$$

with separated boundary conditions

$$y(a) = A, \quad y(b) = B$$

A common and useful way to investigate such problems is to let $y(x, s)$ be the solution of equation (3.3) with initial values $y(a, s) = A$ and $y'(a, s) = s$. For each value of the parameter $s$, the solution $y(x, s)$ satisfies the boundary condition at the end point $x = a$ and extends from $x = a$ to $x = b$. We then ask: For what values of $s$ does $y(x, s)$ satisfy the other boundary condition at $x = b$? – that is, for what values of $s$ is $y(b, s) = B$? If there is a solution $s$ to this algebraic equation then $y(x, s)$ provides a solution of the ODE that satisfies both of the boundary conditions, which is to say that it is a solution of the two-point BVP. By exploiting linearity we can sort out the possibilities easily. Let $u(x)$ be the solution of equation (3.3) defined by the initial conditions $y(a) = A$ and $y'(a) = 0$ and let $v(x)$ be the solution defined by the initial conditions $y(a) = 0$ and $y'(a) = 1$. Linearity implies that the general solution of equation (3.3) is

$$y(x, s) = u(x) + sv(x)$$

The boundary condition

$$B = y(b, s) = u(b) + sv(b)$$

amounts to a linear algebraic equation for the unknown initial slope $s$. The familiar facts of existence and uniqueness of solutions of linear algebraic equations tell us that there is exactly one solution to the BVP when $v(b) \neq 0$, namely the solution of the IVP with

$$s = \frac{B - u(b)}{v(b)}$$

Further, if $v(b) = 0$ then there are infinitely many solutions when $B = u(b)$ and none when $B \neq u(b)$. Existence and uniqueness of solutions of this BVP are clear in principle, but it is obvious that there is a potential for computational difficulties when $v(b) \approx 0$. In extreme circumstances, there is a danger of computing a solution when none exists and even of concluding that no solution exists when one does.

Eigenproblems – more specifically, Sturm–Liouville eigenproblems – are exemplified by the ODE

$$y'' + \lambda y = 0 \tag{3.4}$$

with periodic boundary conditions

$$y(0) = y(\pi), \quad y'(0) = y'(\pi)$$

which are nonseparated, or Dirichlet boundary conditions

$$y(0) = 0, \quad y(\pi) = 0$$

which are separated. For all values of the parameter $\lambda$, the function $y(x) \equiv 0$ is a solution of the BVP, but for some values of $\lambda$ there are also nontrivial solutions. These values are *eigenvalues* and the corresponding nontrivial solutions are *eigenfunctions*. The Sturm–Liouville problem is to compute the eigenvalues and eigenfunctions. This is a nonlinear BVP because the unknown parameter $\lambda$ multiplies the unknown solution $y(x)$. If $y(x)$ is a solution of the BVP, it is easily seen that $\alpha y(x)$ is also a solution for any constant $\alpha$. Accordingly, we need a normalizing condition to specify the solution of interest. For instance, we might require that $y'(0) = 1$. This choice is always valid because $y'(0) = 0$ corresponds to the trivial solution and, if $y'(0) \neq 0$, we can scale $y(x)$ so that $y'(0) = 1$. We can think of a normalizing condition as another boundary condition that is needed to determine the unknown parameter $\lambda$ in addition to the two boundary conditions that determine the solution $y(x)$ of the second-order ODE (3.4).

For values $\lambda > 0$, the solution of the IVP comprising equation (3.4) and initial conditions $y(0) = 0$ and $y'(0) = 1$ is

$$y(x) = \frac{\sin\left(x\sqrt{\lambda}\right)}{\sqrt{\lambda}}$$

If we solve the BVP with Dirichlet boundary conditions, the condition $y(\pi) = 0$ amounts to a nonlinear algebraic equation for the unknown value $\lambda$. Questions of existence and uniqueness of solutions of nonlinear algebraic equations are usually difficult to answer. However, the algebraic equation here is solved easily, and we find that the BVP has a nontrivial solution if and only if the eigenvalue takes one of the values

$$\lambda = k^2, \quad k = 1, 2, \ldots$$

This example shows that, when solving a Sturm–Liouville eigenproblem, we must indicate which of the eigenvalues interests us. There is a large body of theory for the Sturm–Liouville eigenproblem because these BVPs are important in applications and have a very special form. The monograph by Pryce (1993) explains how this theory can be exploited numerically. Some effective codes designed specifically for solving Sturm–Liouville eigenproblems are D02KDF (NAG 2002), SL02F (Marletta & Pryce 1995), SLEIGN (Bailey, Gordon, & Shampine 1978), and SLEDGE (Pruess, Fulton, & Xie 1992). So much is known about this problem and its solution that, when using one of the solvers cited, you can specify precisely what you want. For example, you can say simply that you want the fifth eigenvalue (if they are increasing) and its corresponding eigenfunction. You can also solve Sturm–Liouville eigenproblems with the general-purpose BVP solvers that we study in this chapter, but then you must supply guesses for both the eigenvalue and eigenfunction that interest you. A theoretical result – that, for straightforward problems of this type, the $k$th eigenfunction oscillates $k$ times between the boundary points $a$ and $b$ – can help you make a reasonable guess, but you cannot be certain that the solver will compute the desired eigenvalue and corresponding eigenfunction.

Nonlinearity introduces other complications that are illustrated by the ODE

$$y'' + |y| = 0$$

with separated boundary conditions

$$y(0) = 0, \quad y(b) = B$$

from Bailey, Shampine, & Waltman (1968). If a linear BVP has more than one solution then it has infinitely many, but a nonlinear BVP can have a finite number of solutions. Proceeding as with the linear examples, we find that for any point $b > \pi$ there are exactly two solutions of this BVP for each value $B < 0$. One solution has the form $y(x, s) = s \sinh(x)$. It starts off from the origin with a negative slope $s$ and decreases monotonically to the value $B$ at the boundary point $x = b$. The other solution starts off with a positive slope where it has the form $y(x, s) = s \sin(x)$. This second solution crosses the axis at $x = \pi$ where the form of the solution changes. Thereafter it decreases monotonically to the value $B$ at $x = b$. Figure 3.1 shows an example of this behavior with $b = 4$ and $B = -2$. Much as when solving eigenvalue problems, when solving nonlinear BVPs we must indicate which solution interests us. The MATLAB demonstration program twobvp that produced Figure 3.1 solves the BVP twice, using different guesses for the solution of interest. Specifically, it uses the guess $y(x) \equiv -1$ to compute the solution that is negative on $(0, 4\pi]$ and $y(x) \equiv 1$ to compute the other solution.

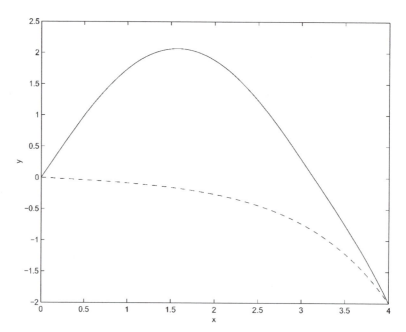

Figure 3.1: A nonlinear BVP with exactly two solutions.

These examples were constructed to make them easy to understand, but other examples in this chapter show that BVPs *modeling physical situations* may not have unique solutions. There are also examples which show that problems involving physical parameters may have solutions only for parameter values in certain ranges. These examples make it clear that, in practice, solving BVPs may well involve an exploration of the existence and uniqueness of solutions of a model. This is quite different from solving IVPs, for which local existence and uniqueness of solutions is assured under mild conditions that are almost always satisfied in practice.

# 3.3 Boundary Conditions

When modeling a physical problem with a BVP, it may not be clear what kinds of boundary conditions to use and where to apply them. In this section we discuss the matter briefly and then turn to the related issue of singularities. It is not at all unusual for BVPs to be singular in the sense that either the ODEs are singular at an end point, or the interval is infinite, or both. The standard theory does not apply to such problems, so if we are to solve them, we must bring to bear all the insight that we can gain from the underlying physical problem and analysis of the mathematical problem. In Sections 3.3.1 and 3.3.2 we use examples to expose some of the important issues. Several examples in Section 3.5 show the details of preparing some realistic singular problems for solution with standard codes, and several exercises outline how to proceed for others.

Generally, the total number of boundary conditions must be equal to the sum of the orders of the ODEs plus the number of unknown parameters. This is entirely analogous to the situation with IVPs. For the first-order systems that most concern us, the number of boundary conditions must be equal to the number of ODEs in the system plus the number of unknown parameters. BVP solvers will fail if you do not supply enough boundary conditions or if you supply too many. Usually the boundary conditions arise from the nature of the problem and usually it is straightforward to determine their type and where to apply them. However, there are occasions when

- It may be difficult to determine one or more of the boundary conditions because all the direct physical constraints on the solution have been imposed and there are still not enough boundary conditions. In such a case you should look for additional boundary conditions arising from the requirement that certain integrals of the system of ODEs should be conserved.
- Some of the boundary conditions may be of a special form. For example, the solution must remain bounded at a singularity or the solution must decay, or decay in a particular way, as the independent variable tends to infinity. In such cases it is usually possible to convert this information into standard boundary conditions, possibly by introducing unknown parameters. A number of examples of both types are discussed in the rest of this section and in Section 3.5.
- There may be too many boundary conditions. Sometimes your knowledge of the underlying problem leads to "knowing" more about the solution than is necessary to specify the BVP. In such cases you should drop those boundary conditions that are consequences of the other boundary conditions and the ODEs. Usually (but not always), such boundary conditions involve higher derivatives of the solution.

## 3.3.1 Boundary Conditions at Singular Points

In this section we discuss singularities at finite points. They often occur in the reduction of a PDE to an ODE by cylindrical or spherical symmetry. For instance, Bratu's equation,

$$\Delta y + e^y = 0$$

is used to model spontaneous combustion. Example 3.5.1 treats the nonsingular case of slab symmetry, but cylindrical and spherical symmetry are even more interesting physically. In these cases the interval is [0, 1] and the ODE is

$$y'' + k\frac{y'}{x} + e^y = 0 \tag{3.5}$$

with $k = 1$ for cylindrical symmetry and $k = 2$ for spherical symmetry.

The obvious difficulty of (3.5) is the behavior of the term $y'/x$ near $x = 0$. We might reasonably expect that there is no real difficulty in solving this BVP because the term is an artifact of the coordinate system and we expect the solution to be smooth there on physical grounds. Indeed, because of symmetry, we must have $y'(0) = 0$, so the term is indeterminate rather than infinite. The standard theory of convergence has been extended to BVPs like this for all the popular numerical methods (de Hoog & Weiss 1976, 1978). For methods that do not evaluate the ODE at $x = 0$, solving a BVP like (3.5) is straightforward. Obviously it is not straightforward for methods that do evaluate at $x = 0$, such as the method of bvp4c. (The bvp4c in version 6.1 of MATLAB cannot solve such BVPs directly, but the one in version 6.5 can.) However, whether a particular solver can deal with this BVP directly is not the point of the example; rather, it is to illustrate how to deal with singularities at finite points.

The idea is first to approximate the solution near the singular point by analytical means. The means employed depend very much on the problem and what you expect of the solution on physical grounds. Often some kind of series or asymptotic approximation is used. We recommend highly the text of Bender & Orszag (1999) because it has many interesting examples showing how to do this. Singularities are much less common when solving IVPs, but they are handled in the same way, so the example and exercises of Section 2.3.4 provide additional examples of dealing with singular points.

For the present example we expect a smooth solution that we can expand in a Taylor series. Taking into account symmetry, this series has the form

$$y(x) = y(0) + \frac{y''(0)}{2}x^2 + \frac{y^{(4)}(0)}{4!}x^4 + \cdots$$

Whatever the assumed form of the solution, we substitute it into the ODE. Using the boundary conditions at $x = 0$, we then determine the coefficients so as to satisfy the equations to as high an order as possible as $x \to 0$. For the example,

$$(y''(0) + \cdots) + \frac{k}{x}(y''(0)x + \cdots) + e^{y(0)+\cdots} = 0$$

Equating to zero the leading (constant) terms, we find that we must take

$$y''(0) = -\frac{e^{y(0)}}{k+1}$$

If we drop the higher-order terms of the Taylor series, this is already enough to provide a good approximation to the solution $y(x)$ on an interval $[0, \delta]$ for a small $\delta > 0$. The derivative of this approximation provides an approximation to $y'(x)$. We now solve the ODEs numerically on the interval $[\delta, b]$ where the equations are not singular. What do we use for boundary conditions at $x = \delta$? Clearly we want the (analytical) approximate solution on $[0, \delta]$ to agree at $x = \delta$ with the numerical solution on $[\delta, b]$. For our example this is

$$y(\delta) = p - \frac{e^p}{2(k+1)}\delta^2$$

$$y'(\delta) = -\frac{e^p}{k+1}\delta$$

Here we have written $p = y(0)$ to emphasize that it is an unknown parameter that must be determined as a part of solving the BVP. This example is typical of the way that unknown parameters are introduced to deal with singularities.

A boundary condition tells us how the solution of interest behaves as it approaches an end point. This is not always as simple as approaching a given value. To describe more complex behavior, we use the standard notation

$$f(x) \sim F(x)$$

as $x \to x_0$ to mean

$$\lim_{x \to x_0} \frac{f(x)}{F(x)} = 1$$

This is read as "$f(x)$ is asymptotic to $F(x)$ as $x$ approaches $x_0$". The boundary condition $y'(0) = 0$ implies that the solution of Bratu's problem tends to a constant as $x \to 0$, but other boundary conditions are also possible for this ODE. To explore this, we'll investigate solutions that are not bounded at the origin. We begin with $k = 1$. Supposing that derivatives of a solution grow more rapidly than the solution itself as $x \to 0$, we might try approximating the ODE with

$$0 = y'' + \frac{y'}{x} + e^y \sim y'' + \frac{y'}{x}$$

Solving the approximating equation, we find that

$$y(x) \sim b + a \log(x) = b + \log(x^a)$$

for constants $a$ and $b$. We now substitute this analytical approximation into the ODE to see when it might be valid:

$$y'' + \frac{y'}{x} + e^y \sim -\frac{a}{x^2} + \frac{a}{x^2} + x^a e^b$$

To satisfy the ODE asymptotically, we must have $a > 0$. Boundary conditions like $y(x) \sim \log(x)$ as $x \to 0$ arise, for example, when solving potential problems with a line charge at the origin. We see now that the ODE has solutions that behave in this way, so this is a legitimate boundary condition. Similarly, when $k = 2$,

$$y(x) \sim b + ax^{-1}$$

and it is necessary that $a < 0$. Boundary conditions like $y(x) \sim -x^{-1}$ arise, for example, when solving potential problems with a point charge at the origin. It is no harder to solve Bratu's equation with the boundary condition $y(x) \sim \log(x)$ than with the symmetry condition: we use the analytical approximations $y(x) \approx \log(x)$ and $y'(x) \approx x^{-1}$ on $(0, \delta]$ for some small $\delta$ and solve numerically the ODEs on $[\delta, 1]$ with boundary value $y(\delta) = \log(\delta)$. Note that with this particular boundary condition there is no unknown parameter $y(0)$ as there was with the symmetry condition.

It can be more difficult to determine the behavior of solutions at a singular point. Bender & Orszag (1999, pp. 170–2) discuss the solution of the ODE

$$yy'' = -1 \qquad\qquad (3.6)$$

with boundary conditions $y(0) = 0$ and $y(1) = 0$. If we are to have $y(x) \to 0$ as $x$ tends to 0 and 1, then the ODE requires that $y''(x)$ be unbounded there. Clearly we must supply a solver with more than just limit values for $y(x)$. We follow Bender and Orszag in developing asymptotic approximations to $y(x)$. The solution is symmetric about $x = 0.5$, so we need only investigate the behavior of the solution near the origin. We must expect (at least) two solutions of the BVP because, if $y(x)$ is a solution, then so is $-y(x)$. The solution vanishes at the origin and has derivatives that are infinite, so let's try something simple like

$$y(x) \sim ax^b$$

for constants $a$ and $b$. Substituting into the ODE, we find that

$$-1 = y(x)y''(x) \sim (ax^b)(b(b-1)x^{b-2}) = ab(b-1)x^{2b-2}$$

If the power $2b - 2 < 0$, then the right-hand side does not have a limit as $x \to 0$. If $2b - 2 > 0$, there is a limit but it is zero. If $2b - 2 = 0$, the right-hand side is identically zero. Thus we see that there is no choice that allows us to satisfy the ODE, even asymptotically. This tells us that the form we have assumed for the solution is wrong. After some experimentation, we might be led to try the form

$$y(x) \sim ax(-\log(x))^b$$

Substituting this form into the ODE, we find after some manipulation that

$$-1 = y(x)y''(x) \sim -a^2 b(-\log(x))^{2b-1}[1 - (b-1)(-\log(x))^{-1}]$$

If the right-hand side is to have a limit as $x \to 0$ then we must have the power $2b - 1 = 0$, hence $b = 0.5$. With this we pass to the limit and find that $a = \pm\sqrt{2}$. It appears that there are exactly two solutions. The one that is positive for positive $x$ is $y(x) \sim x\sqrt{-2\log(x)}$. In writing this second-order equation as a first-order system for its numerical solution, we introduce $y'(x)$ as an unknown. Because it is infinite at the origin, we need to know how it behaves there if we are to solve the BVP. On the other hand, all positive solutions of the ODE that vanish at the origin behave in the same way to leading order, so it is not necessary to introduce an unknown parameter if we are interested only in modest accuracy. Exercise 3.4 asks you to see for yourself that, once you prepare the BVP with this information about how the solution behaves near the singular point, you can solve it easily.

For further examples with a physical origin, you should read now the portions of the following exercises that are devoted to the behavior of the solutions. See particularly Exercise 3.2. Example 3.5.4 and Exercise 3.9 take up singular boundary conditions at the origin, but they have the further complication of an infinite range. After you have learned how to use the MATLAB BVP solver, you can return to the exercises and solve the BVPs numerically.

■ **EXERCISE 3.1**

If the solution of a BVP behaves well at a singular point, it may not be necessary to use an analytical approximation in the neighborhood of the point. To illustrate this we consider a Sturm–Liouville eigenproblem called Latzko's equation,

$$\frac{d}{dt}\left((1 - x^7)\frac{dy}{dx}\right) + \lambda x^7 y = 0$$

with boundary conditions $y(0) = 0$ and $y(1)$ finite. A number of authors have solved this problem numerically because of its physical significance and interesting singularity. Scott (1973, p. 153) collects approximations to the first three eigenvalues and compares them to his approximations: 8.728, 152.45, and 435.2. To write this equation as a first-order system, it is convenient to use the unknowns $y(x)$ and $v(x) = (1 - x^7)y'(x)$ and so obtain

$$y' = \frac{v}{1 - x^7}$$
$$v' = -\lambda x^7 y$$

This is an eigenvalue problem, so we must specify a normalizing condition (e.g., $y(1) = 1$) for the solution. The differential equation is singular at $x = 1$. If we are looking for a solution with $y'(1)$ finite, the first equation of the system states that we must have $v(1) = 0$. Using the second equation, L'Hôpital's rule then says that

$$y'(1) = \lim_{x \to 1}\frac{v(x)}{1 - x^7} = \lim_{x \to 1}\frac{v'(x)}{-7x^6} = \lim_{x \to 1}\frac{-\lambda x^7 y(x)}{-7x^6} = \frac{\lambda}{7}$$

We can solve the ODEs for $y(x)$ and $v(x)$ subject to the boundary conditions

$$y(0) = 0, \quad y(1) = 1, \quad v(1) = 0$$

provided that we evaluate them properly at $x = 1$. When solving a BVP with an unknown parameter, bvp4c calls the function for evaluating the ODEs with a mesh point $x$ and the current approximations to $y(x)$ and $\lambda$. If $x = 1$, this function must return the limit value for $y'(1)$. You must code for this because this solver always evaluates the ODEs at the end points of the interval. With a smooth solution, the solver will not need to evaluate the ODEs at points $x$ so close to the singular point that there are difficulties evaluating $y'(x)$, difficulties that we avoid for other kinds of problems by using analytical approximations near the singularity. Of course, the guess for $v(x)$ should have the proper behavior near $x = 1$. This is easily accomplished by using the guess for $y(x)$ to form $(1 - x^7)y'(x)$ as a guess for $v(x)$.

A different way to perform this analysis is to look for a solution $y(x) \sim 1 + c(x - 1)$ as $x \to 1$. If there is a solution of this form, then $y'(x) \sim c$. Because $y'(1)$ is finite for a solution of this form, it follows that $v(1) = 0$. Determine the constant $c$ by substituting this assumed form into the ODE for $y(x)$. In this it will be helpful to show that $(1 - x^7) \sim 7(1 - x)$ near $x = 1$. If the function for evaluating the ODEs is coded properly, then it is easy enough to solve this eigenproblem with bvp4c. What is not easy is to compute specific eigenvalues and eigenfunctions. This is because it is not clear which solution you will compute with given guesses for $y(x)$ and $\lambda$, or even whether you will compute a solution. As mentioned in the text, there are codes for the Sturm–Liouville eigenproblem that deal with this by bringing to bear a deep understanding of the theory of such problems. Using bvp4c, try to confirm one or more of Scott's approximate eigenvalues. We found that the guess $y(x) \approx x \sin(2.5\pi x)$ and an initial mesh of linspace(0,1,10) yielded convergence to different eigenvalues for guesses $\lambda = 10, 100, 500, 1000, 5000$, and 10,000; these computed eigenvalues include the first three. This behavior is somewhat surprising since – with different guesses for $y(x)$ or even a different initial mesh – these guesses for $\lambda$ often do not result in different eigenvalues.

## ■ EXERCISE 3.2

The first example of the documentation for the BVP code D02HBF in the NAG library (NAG 2002) is the singular problem

$$2xy'' + y' = y^3, \quad y(0) = \frac{1}{10}, \quad y(16) = \frac{1}{6}$$

Look for a solution that behaves like

$$y(x) \sim \alpha + px^\beta + \gamma x$$

as $x \to 0$. In this assume that $0 < \beta < 1$. Determine $\alpha$, $\beta$, and $\gamma$ from the ODE and the boundary condition at $x = 0$. The coefficient $p$ is an unknown parameter. A solution that behaves in this way has a first derivative that is not bounded as $x \to 0$. You are not asked to solve this BVP numerically, but to do so you would use these asymptotic approximations to $y(x)$ and $y'(x)$ on an interval $(0, d]$ for a small $d$ and solve numerically a BVP on $[d, 16]$. Requiring that the numerical solution agree at $d$ with the asymptotic solution provides boundary conditions for the BVP. It is necessary to require continuity in both $y(x)$ and $y'(x)$ so as to have enough conditions to determine the unknown parameter $p$. It would be reasonable to use the asymptotic approximations along with a guess for $p$ as guesses for the solution on $[d, 16]$.

### ■ EXERCISE 3.3

Section 6.2 of Keller (1992) discusses the numerical solution of a model of the steady concentration of a substrate in an enzyme-catalyzed reaction with Michaelis–Menten kinetics. A spherical region is considered and the PDE is reduced to an ODE by symmetry. The equation

$$y'' + 2\frac{y'}{x} = \frac{y}{\varepsilon(y + k)}$$

involves two physical parameters, $\varepsilon$ and $k$. The boundary conditions are $y(1) = 1$ and the symmetry condition $y'(0) = 0$. Once you have learned how to use bvp4c, solve this problem for the parameter values $\varepsilon = 0.1$ and $k = 0.1$. Approximate $y(x)$ on the interval $[0, d]$ with a few terms of a Taylor series expansion and solve the BVP numerically on $[d, 1]$. To plot the solution on all of $[0, 1]$, augment the numerical solution on the interval $[d, 1]$ with the values at $x = 0$ provided by the unknown parameter $p = y(0)$ and the initial derivative $y'(0) = 0$. Two terms of a Taylor series for $y(x)$ and one for $y'(x)$ are satisfactory when $d = 0.001$. You will need to communicate $d$ and the parameters $\varepsilon$ and $k$ to your functions for evaluating the ODEs and boundary conditions. You could hard code them, pass them as global variables, or pass them as optional input to bvp4c. If you choose the last possibility, keep in mind that these optional arguments must follow the unknown parameter $p$ in the various call lists. Also, if you do not set any options, you will need to use [] as a placeholder in this argument that precedes unknown parameters in the call list of bvp4c.

### ■ EXERCISE 3.4

Once you have learned how to use bvp4c, solve the BVP

$$yy'' = -1, \quad y(0) = 0, \ y'(0.5) = 0$$

The text shows that $y(x) \sim x\sqrt{-2\log(x)}$ as $x \to 0$. It will be convenient to code a subfunction to evaluate the approximation $v(x) = x\sqrt{-2\log(x)}$. Move the boundary condition from the singular point at the origin to, say, $d = 0.001$ by requiring that $y(d) = v(d)$.

Compute $y(x)$ on $[d, 0.5]$ using `bvp4c` with default tolerances and guesses of $y(x) \approx x(1-x)$ and $y'(x) \approx 1 - 2x$. Corresponding to Figure 4.11 of Bender & Orszag (1999), plot both $y(x)$ and $v(x)$ with `axis([0 0.5 0 0.5])`. In this you should augment the array for $y(x)$ with $y(0) = 0$ and the array for $v(x)$ with $v(0) = 0$.

### 3.3.2  Boundary Conditions at Infinity

Whether a BVP is well-posed depends on the nature of solutions of the ODEs and on the boundary conditions. Appropriate boundary conditions may be obvious from physical reasoning but, when they are not, you need to be aware of the issues. This matter is always important for problems set on infinite ranges, but it is also important when the interval is "large". After we understand better what kinds of boundary conditions might be imposed at infinity, we consider some examples that show how to prepare problems for solution with codes that require a finite interval. This is very much like the way we deal with a singularity at a finite end point.

The equation

$$y''' + 2y'' - y' - 2y = 0 \qquad (3.7)$$

is illuminating. Its general solution is

$$y(x) = Ae^{x} + Be^{-x} + Ce^{-2x}$$

Notice that there are three components of the solution, two that decay as $x$ increases and one that grows. Suppose that we solve this equation on the interval $[0, +\infty)$ with boundary conditions

$$y(0) = 1, \quad y'(0) = 1, \quad y(+\infty) = 0$$

The last boundary condition,

$$y(\infty) = 0 = A \times \infty + B \times 0 + C \times 0$$

implies that $A = 0$. The other two conditions,

$$y(0) = 1 = A \times 1 + B \times 1 + C \times 1$$
$$y'(0) = 1 = A \times 1 - B \times 1 - 2C \times 1$$

imply that $B = 3$ and $C = -2$. From this we see that there is a solution of this BVP and only one solution. On the other hand, if the boundary conditions are

$$y(0) = 1, \quad y(+\infty) = 0, \quad y'(+\infty) = 0 \qquad (3.8)$$

then the boundary condition $y(\infty) = 0$ again implies that $A = 0$, but now the last condition,

$$y'(\infty) = 0 = A \times \infty - B \times 0 - 2C \times 0$$

places no constraint on the coefficients. The remaining boundary condition,

$$y(0) = 1 = A \times 1 + B \times 1 + C \times 1$$

tells us only that $C = 1 - B$, so any value of $B$ results in a solution; that is, this BVP has infinitely many solutions.

If a BVP is not well-posed with boundary conditions at $b = +\infty$, it is natural to expect numerical difficulties when they are imposed at a finite point $b \gg +1$. Suppose then that we solve equation (3.7) with boundary conditions

$$y(0) = 1, \quad y(b) = 0, \quad y'(b) = 0$$

replacing (3.8). For large values of $b$, the system of linear equations for the coefficients $A$, $B$, and $C$ in the general solution is extremely ill-conditioned. Indeed, even for $b$ as small as 20, an attempt to invert the matrix results in (a) a warning from MATLAB that the matrix is close to singular and (b) an estimate of $5 \cdot 10^{17} = 1/\texttt{RCOND}$ for its condition number. The essence of the matter is that the two solutions of the ODE that decay exponentially fast as $x$ increases cannot be distinguished numerically by their values at $x = b$; hence the linear system for their coefficients in the linear combination that forms the solution of the BVP is numerically singular. The third solution, $e^x$, decays exponentially fast as $x$ decreases, leading to an ill-conditioned linear system if $b$ is large and the boundary condition involves the value of this solution at $x = 0$. Exercise 3.5 considers a related problem.

Generalizing this last example to a system of linear ODEs with constant coefficients,

$$y' = Jy + q \tag{3.9}$$

gives more insight. For the sake of simplicity, let us suppose that the Jacobian matrix $J$ is nonsingular. This implies that $p(x) = -J^{-1}q$ is a constant particular solution of the ODE. Further assume that $J$ has a complete set of eigenvectors $\{v^j\}$ with corresponding eigenvalues $\{\lambda_j\}$. This implies that there are constants $\alpha_j$ such that

$$y(a) - p(a) = \sum \alpha_j v^j$$

It is then verified easily that the general solution of the system of ODEs is

$$y(x) = p(x) + \sum \alpha_j e^{\lambda_j (x-a)} v^j$$

We see from this that $y(x)$ is finite on $[a, +\infty)$ only when the boundary conditions at $x = a$ imply that $\alpha_m = 0$ for all $m$ with $\text{Re}(\lambda_m) > 0$. Correspondingly, the BVP is well-conditioned for $b \gg a$ only when the boundary conditions at $x = a$ exclude the terms that grow exponentially fast as $x$ increases. Similarly, using the expansion

$$y(b) - p(b) = \sum \beta_j v^j$$

we see that if $b \gg a$ then we must have boundary conditions at $x = b$ that imply $\beta_m = 0$ for any value $m$ such that $\text{Re}(\lambda_m) < 0$ if the BVP is to be well-conditioned. Roughly speaking, components that decay rapidly from left to right must be determined by the boundary conditions at the left end of the interval, and components that grow rapidly in this direction must be determined by the boundary conditions at the right end.

These linear, constant-coefficient problems are rather special, but they provide valuable insight. More generally we recognize that the boundary conditions restrict the various kinds of behavior that solutions can have. If there are solutions of the ODE that approach one another very quickly as $x$ increases from $a$ to $b$, then for $b \gg a$ a code will not be able to distinguish the solutions numerically when applying the boundary conditions at $x = b$. Because of this, these solutions must be distinguished by boundary conditions at the point $x = a$ rather than at $x = b$. There is a *dichotomy* of the solution space of the ODE that must be reflected in the placement of the boundary conditions. This is easy enough to sort out for linear, constant-coefficient ODEs, but for more general ODEs some careful physical reasoning is normally needed to decide where and what kind of boundary conditions are appropriate. Ascher, Mattheij, & Russell (1995) provide more details and examples that illuminate this issue.

An interesting and informative example is provided by traveling wave solutions of Fisher's equation. The significance of these solutions and a thorough analysis of the problem is found in Section 11.2 of Murray (1993). A wave traveling at speed $c$ has the form $u(x, t) = U(z)$. Here $z = x - ct$ and $U(z)$ satisfies the ODE

$$U'' + cU' + U(1 - U) = 0 \tag{3.10}$$

Obviously this equation has two trivial steady states, $U(z) \equiv 1$ and $U(z) \equiv 0$. A typical wavefront solution has

$$U(-\infty) = 1, \quad U(+\infty) = 0$$

(Together these boundary conditions preclude the steady-state solutions.) For a given value of the wave speed $c$, it appears that we have the ODE and boundary conditions that we need to define properly a BVP, but it is easy to see that we do not: there are an infinite number of solutions, because if $U(z)$ is a solution of this BVP then so is $U(z + \gamma)$ for any constant $\gamma$. It is tempting to choose some large number $Z$, specify the boundary conditions $U(-Z) = 1$ and $U(Z) = 0$ to replace the corresponding boundary conditions at $-\infty$ and $+\infty$ (respectively), and then call upon our favorite BVP solver. This approach is not likely to succeed because the BVP has an infinite number of solutions and we have provided no information that distinguishes a specific one. Murray studies this BVP in the phase plane $(U, U')$. There are two singular points $(0, 0)$ and $(1, 0)$ that correspond to the steady states. A linear stability analysis shows that, if $c \geq 2$, then the critical point $(0, 0)$ is a stable node and the critical point $(1, 0)$ is a saddle point. By examining trajectories in the phase plane, Murray argues that if $c \geq 2$ then there exists a solution of the BVP such that, on approach to the point $(0, 0)$,

$$U'(z) \sim \beta U(z)$$

with $\beta = \left(-c + \sqrt{c^2 - 4}\right)/2$ and, on approach to the point $(1, 0)$,

$$(U(z) - 1)' \sim \alpha(U(z) - 1)$$

with $\alpha = \left(-c + \sqrt{c^2 + 4}\right)/2$.

Linear stability analysis in a phase plane amounts to approximating the equations by an ODE with constant coefficients – that is, an ODE of the form (3.9). We can proceed in a more direct way. We learned that with assumptions that exclude degenerate cases, the solution of (3.9) consists of a constant particular solution plus a linear combination of solutions of the homogeneous problem that are exponentials. So let's look for a solution of this form that approaches the steady state $(1, 0)$ as $z \to -\infty$. Specifically, we look for a solution of the form

$$U(z) \sim 1 + pe^{\alpha z}$$

and, correspondingly,

$$U'(z) \sim \alpha pe^{\alpha z}$$

If the point $(U(z), U'(z))$ is to approach this steady state, we must have $\mathrm{Re}(\alpha) > 0$. Substituting into the ODE, we find that we want

$$U'' + cU' = (U - 1)U$$
$$\alpha^2 pe^{\alpha z} + c\alpha pe^{\alpha z} \sim pe^{\alpha z} + p^2 e^{2\alpha z}$$

Dividing out nonzero common factors, we need a value $\alpha$ for which

$$\alpha^2 + c\alpha \sim 1 + pe^{\alpha z} \sim 1$$

as $z \to -\infty$. From this we find that $\alpha = \left(-c \pm \sqrt{c^2 + 4}\right)/2$. The two values reflect the fact that this steady state is a saddle point in the phase plane. Only the positive value $\alpha = \left(-c + \sqrt{c^2 + 4}\right)/2$ provides the behavior that we require. Now we must select a boundary condition that imposes the correct kind of behavior as $z \to -\infty$. We have worked out the decay rate, but we do not know the factor $p$ because it is specific to the solution we seek. Either we must introduce $p$ as an unknown parameter or we must formulate the condition in a way such that $p$ does not appear. A simple way to do this is to require that

$$\frac{U'(z)}{U(z) - 1} \to \alpha$$

That is, we choose a number $Z \gg 1$ and require that

$$\frac{U'(-Z)}{U(-Z) - 1} - \alpha = 0$$

as our left boundary condition.

For illustrative purposes, we proceed a little differently in working out the behavior of $U(z)$ as $z \to +\infty$. Because $U(z)$ tends to zero, we can approximate $U(z)(1 - U(z))$ by $U(z)$ for large $z$. Accordingly, we can approximate (3.10) by the ODE

$$U'' + cU' + U = 0$$

Solving this ODE yields

$$U(z) \sim q e^{\beta z}$$

with $\beta$ a root of $\beta^2 + c\beta + 1 = 0$. Both roots of this quadratic are negative, and it is not immediately obvious which should be chosen. A more detailed analysis by Murray indicates that we must use $\beta = \left(-c + \sqrt{c^2 - 4}\right)/2$. Essentially, we work with the more slowly decaying component of the solution, since this guarantees that the faster decaying component has effectively vanished at the boundary point. Again there is an issue of the factor $q$ and the form of the boundary condition. Recall that if $U(z)$ is a solution of the ODE then so is $U(z + \gamma)$. To deal with this, it is convenient here to choose a value for $q$ so as to compute a specific solution. Like Murray, we choose $q = 1$ and impose the right boundary condition

$$\frac{U(Z)}{e^{\beta Z}} - 1 = 0$$

This problem is considered further in Example 3.5.5; see also Exercises 3.25 and 3.26.

This discussion may have left the impression that, when the BVP is posed on an infinite range, solutions always decay at an exponential rate. However, there are other kinds of behavior that are common in practice. As a simple example, consider the equation

$$y' = \lambda x y$$

with boundary conditions $y(0) = 1$ and $y(x) \sim e^{-x^2}$ as $x \to \infty$. Here $\lambda$ is an unknown parameter that is found analytically by solving the ODE with initial value $y(0) = 1$ and then imposing the boundary condition at infinity. In this way we find easily that the solution is

$$\lambda = -2, \quad y(x) = e^{-x^2}$$

When solving problems set on an infinite range, we often replace boundary conditions at infinity by boundary conditions at some large $b$. Because many problems set on an infinite range have solutions that decay exponentially fast, our experience may lead us to choose a value that is too large when solutions actually decay much faster, as in this example. Choosing a $b$ that is much too large can cause a BVP code to fail because it cannot distinguish numerically the various kinds of solution at $x = b$.

At the other extreme are BVPs with solutions that decay algebraically rather than exponentially. As a simple example that we use to make another point, consider the equation

$$y' = -\lambda y^2$$

with boundary conditions $y(0) = 1$ and $y(+\infty) = 0$. Again $\lambda$ is an unknown parameter. The analytical solution of the ODE that satisfies $y(0) = 1$ is

$$y(x) = \frac{1}{\lambda x + 1}$$

This solution satisfies the boundary condition at infinity for any $\lambda > 0$, which is to say that the BVP has infinitely many solutions. Though artificial, this example makes the point that we may need to provide more information about how the solution of a singular problem behaves if we are to have a well-posed problem. Here we might specify the boundary condition $y(x) \sim x^{-1}$ as $x \to +\infty$ and then find that there is a unique solution with $\lambda = 1$. With an algebraic rate of decay like this, it is generally necessary to use what seems like quite a large value of $b$ to represent infinity. To appreciate this, suppose that we replace the boundary condition at infinity with $y(b) = 1/b$ for some large $b$. It is easily found that the solution on this finite interval is

$$\lambda_b = 1 - \frac{1}{b}, \quad y_b(x) = \frac{1}{x + 1 - x/b}$$

Clearly we would need to use quite a large $b$ for the solution on the interval $[0, b]$ to agree (to even modest accuracy) with the solution of the problem set on the interval $[0, \infty)$.

For further examples with a physical origin, you should read now the portions of the following exercises that are devoted to the behavior of the solutions. After you have learned how to use the MATLAB BVP solver, you can return to the exercises and solve the BVPs numerically.

### ■ EXERCISE 3.5

Show that the general solution of the ODE

$$y''' - 2y'' - y' + 2y = 0$$

is

$$y(x) = Ae^x + Be^{2x} + Ce^{-x}$$

Show that, with boundary conditions

$$y(0) = 1, \quad y(+\infty) = 0, \quad y'(+\infty) = 0$$

there is a unique solution to the BVP. Show that with boundary conditions

$$y(0) = 1, \quad y'(0) = 1, \quad y(+\infty) = 0$$

the BVP does not have a solution. Investigate numerically the condition of the linear system that determines the constants $A$, $B$, and $C$ when the boundary conditions are

$$y(0) = 1, \quad y'(0) = 1, \quad y(20) = 0$$

■ **EXERCISE 3.6**

Murphy (1965) extends the classical Falkner–Skan similarity solutions for laminar incompressible boundary layer flows to flows over curved surfaces. He derives a BVP consisting of the ODE

$$f'''' + (\Omega + f)f''' + \Omega ff'' = \gamma[f'f'' + \Omega(f')^2]$$

and boundary conditions

$$f(0) = 0, \quad f'(0) = 0$$

and, as $x \to \infty$,

$$f'(x) \sim e^{-\Omega x}, \quad f''(x) \sim -\Omega e^{-\Omega x}$$

Here $\Omega$ is a positive curvature parameter and $\gamma$ is a constant related to the pressure gradient. Look for solutions of the ODE that behave like

$$f(x) \sim \delta + \rho e^{-\lambda x}$$

as $x \to \infty$. That is, look for values of the constants $\delta$, $\rho$, and $\lambda > 0$ such that $\delta + \rho e^{-\lambda x}$ satisfies the ODE as $x \to \infty$. You should find that $\lambda = \Omega$ is one possibility. By choosing the constants properly, show that there are solutions that satisfy the boundary conditions at infinity. With this we see that the boundary conditions specified by Murphy are consistent with the behavior of solutions of the ODE. As a by-product of your analysis, find an *exact* solution of the ODE that satisfies the two boundary conditions at infinity and one of the boundary conditions at the origin. This solution would be useful as a guess when solving the BVP numerically.

■ **EXERCISE 3.7**

Ames & Lohner (1981) study models for the transport, reaction, and dissipation of pollutants in rivers. One model gives rise to a system of three first-order PDEs in one space variable $x$ and time $t$. By looking for traveling wave solutions that depend only on the variable $z = x - t$, they reduce the PDEs to the ODEs

$$f'' = \beta g f, \quad g'' = -\beta g h, \quad h'' = \lambda \beta g h$$

Here $f$ represents a pollutant, $g$ bacteria, and $h$ carbon; the physical parameters $\beta$ and $\lambda$ are constants. After showing that the equations for $g$ and $h$ imply that

$$g(z) = E - \frac{h(z)}{\lambda}$$

where $E$ is a given value for $g(\infty)$, they reduce the system of ODEs to

$$h'' = \lambda \beta \left( E - \frac{h}{\lambda} \right) h, \qquad f'' = \beta \left( E - \frac{h}{\lambda} \right) f$$

These equations are to be solved subject to boundary conditions

$$h(0) = 1, \quad f(0) = 1, \quad h(\infty) = 0, \quad f(\infty) = 0$$

Ames and Lohner consider a number of choices of parameters. For $\beta = 10$, $\lambda = 10$, and $E = 1$, they derive an upper bound for $f(0.1)$ of 0.730 and a lower bound of 0.729. For these values of the parameters, argue that $h(z)$ is asymptotically a multiple of $e^{-10z}$ and that $f(z)$ is a multiple of $e^{-\sqrt{10}z}$. In view of this behavior, solve the BVP by replacing the boundary conditions at infinity with the boundary conditions $h(Z) = 0$ and $f(Z) = 0$ for some finite value $Z$ of modest size. Several values should be tried (e.g., $Z = 2$, 3, and 4) in order to gain confidence in your solution. A common way to proceed when solving BVPs on infinite intervals is to guess a value for $Z$. If you cannot solve the BVP, reduce $Z$ and try again. If you can solve the BVP, inspect the solution to see whether it exhibits the desired behavior well before the end of the interval. If not, increase $Z$ and try again. For this problem you could use the asymptotic expressions for $h(z)$ and $f(z)$ as guesses for the first interval, but you will find that simple constant guesses are satisfactory. Check your numerical solution for $f(z)$ against the bounds derived by Ames and Lohner.

■ **EXERCISE 3.8**

Example 7.3 of Bailey et al. (1968) considers a similarity solution for the unsteady flow of a gas through a semi-infinite porous medium initially filled with gas at a uniform pressure. The BVP is

$$w''(z) + \frac{2z}{\sqrt{1 - \alpha w(z)}} w'(z) = 0$$

with boundary conditions

$$w(0) = 1, \quad w(+\infty) = 0$$

A range of values of the parameter $0 < \alpha < 1$ is considered when this problem is solved numerically in Example 8.4 of Bailey et al. (1968). Solve this problem for $\alpha = 0.8$. Although the examples of this chapter emphasize analyzing the behavior of solutions near a singular point, it is often the case that problems on infinite intervals can be solved in a straightforward way. Solve this problem by replacing the boundary condition at infinity with the boundary condition $w(Z) = 0$ for some finite value $Z$. Several values should be tried (e.g., $Z = 2$, 3, and 4) in order to gain confidence in your solution. Your guess for $w(z)$ should respect the physical requirements that $0 \le w(z) \le 1$. There should then be no difficulty in forming the square root in the ODE.

To understand better what is going on, we first observe that because $w(\infty) = 0$, the ODE is approximately $w''(z) + 2zw'(z) = 0$ for large $z$. Solving this approximating equation, we find that

$$w'(z) \sim \beta e^{-z^2}$$

Integrating and imposing the boundary condition at infinity, we then find that

$$w(z) \sim -\beta \int_z^\infty e^{-t^2}\, dt = \left( -\frac{\beta \sqrt{\pi}}{2} \right) \text{erfc}(z)$$

The standard asymptotic representation of the complementary error function,

$$\text{erfc}(z) \sim \frac{e^{-z^2}}{z\sqrt{\pi}}$$

shows that $w(z)$ approaches its boundary value $w(\infty) = 0$ very quickly. That is why we can impose the numerical boundary condition $w(Z) = 0$ at what seems like a very small value of $Z$. In fact, we must use a small value of $Z$ because the solver cannot distinguish $w(z)$ from the identically zero solution of the ODE when $z$ is large. Sometimes we must supply more information about how a solution behaves near a singular point if we are to compute a numerical solution at all. For this BVP, we could do that by treating the constant $\beta$ as an unknown parameter and imposing the two numerical boundary conditions

$$w'(Z) = \beta e^{-Z^2}, \qquad w(Z) = \left( -\frac{\beta \sqrt{\pi}}{2} \right) \text{erfc}(Z)$$

instead of $w(Z) = 0$. The analytical approximations to $w(z)$ and $w'(z)$ are accurate only for large $z$, but they provide reasonable guesses for the solver for all $z$.

■ **EXERCISE 3.9**

The Thomas–Fermi equation,

$$y'' = x^{-1/2} y^{3/2}$$

is to be solved with boundary conditions

$$y(0) = 1, \quad y(+\infty) = 0$$

This BVP arises in a semiclassical description of the charge density in atoms of high atomic number. There are difficulties at both end points; these difficulties are discussed at length in Davis (1962) and in Bender & Orszag (1999). Davis discusses series solutions for $y(x)$ as $x \to 0$. It is clear that there are fractional powers in the series. That is because, with $y(0) = 1$, the ODE requires that $y'' \sim x^{-1/2}$ as $x \to 0$ and hence that there be a term $\frac{4}{3} x^{3/2}$ in the series for $y(x)$. Of course, there must also be lower-order terms so as to satisfy the boundary condition at $x = 0$. It is natural then to try for a solution of the form

$$y(x) = 1 + px + \tfrac{4}{3} x^{3/2} + bx^2 + cx^{5/2} + \cdots$$

The manipulations will be easier if you write the equation as

$$x(y'')^2 = y^3$$

For this expansion, verify that $p$ is a free parameter, $b = 0$, and $c = 2p/5$. Bender and Orszag discuss the asymptotic behavior of $y(x)$ as $x \to \infty$. Verify that trying a solution of the form $y(x) \sim ax^\alpha$ yields $y_0(x) = 144 x^{-3}$ as an exact solution of the ODE that satisfies the boundary condition at infinity. To study the behavior of the general solution near this particular solution, first write $y(x) = y_0(x) + \varepsilon(x)$. Show that, for functions $\varepsilon(x)$ that are small compared to $y_0(x)$,

$$y^{3/2}(x) \approx y_0^{3/2}(x) + \tfrac{3}{2} y_0^{1/2}(x) \varepsilon(x)$$

and then that the correction $\varepsilon(x)$ is obtained approximately as a solution of $\varepsilon''(x) = 18 x^{-2} \varepsilon(x)$. Look for solutions $\varepsilon(x)$ of the form $cx^\gamma$. Discarding solutions that do not tend to zero as $x \to \infty$, conclude that solutions of the ODE that are near $y_0(x)$ and satisfy the boundary condition at infinity have the form $y(x) \sim 144 x^{-3} + cx^\gamma$ for $\gamma = (1 - \sqrt{73})/2 \approx -3.772$ and an arbitrary constant $c$.

Move the boundary condition from the origin to a small value of $d > 0$ by matching $y(d)$ and $y'(d)$ with values computed using the series approximation and treating $p$ as an unknown parameter. Move the boundary condition at infinity to a large value of $D$ by matching $y(D)$ to $y_0(D)$. Because solutions decay at only an algebraic rate, a relatively large value of $D$ is needed. Solve the BVP with default tolerances and several choices of $d$ and $D$. You might, for example, use intervals $[1, 20]$, $[0.1, 40]$, and $[0.01, 60]$. As a guess for the solution on $[d, D]$, you might use

$$y(x) \approx 1, \quad y'(x) \approx 0$$

for $x \in [d, 1]$ and

$$y(x) \approx 144x^{-3}, \quad y'(x) \approx -432x^{-4}$$

for $x \in (1, D]$. After solving the problem for one choice of $[d, D]$, you could use the capability of `bvpinit` of extrapolating the solution on one interval to form a guess for a longer interval. You should code $x^{-1/2}y^{3/2}$ as

```
max(y(1),0)^(3/2) / sqrt(x)
```

because intermediate approximations to $y(x)$ might be negative, resulting in complex values of the fractional power. Plot together the solutions computed for different intervals to gain some confidence that your last interval provides an acceptable solution. Figure 4.10 of Bender & Orszag (1999) shows some results of solving the BVP with a shooting method. Augment your last solution with $y(0) = 1$ and plot it using `axis[0 15 0 1]` so as to confirm this figure. The parameter $p = y'(0)$, so compare the value for $p$ that you compute to the value $p = -1.588$ obtained in Bender & Orszag (1999) and Davis (1962).

# 3.4 Numerical Methods for BVPs

The theoretical approach to BVPs of Section 3.2 is based on the solution of IVPs and the solution of nonlinear algebraic equations. Because there are effective programs for both tasks, it is natural to combine them to obtain a program for solving BVPs. A method of this kind is called a *shooting method*. Because it appears so straightforward to use quality numerical tools for the solution of BVPs by shooting, it is perhaps surprising that the most popular solvers are *not* shooting codes. The basic difficulty with shooting is that a well-posed BVP can require the integration of IVPs that are unstable. That is, the solution of a BVP can be insensitive to changes in its boundary values, yet the solutions of the IVPs arising in the shooting method for solving the BVP are sensitive to changes in the initial values. The simple example

$$y'' - 100y = 0$$

with boundary conditions $y(0) = 1$ and $y(10) = B$ makes the point. Shooting (from left to right) involves solving the IVP with initial values $y(0) = 1$ and $y'(0) = s$. The analytical solution of this IVP is

$$y(x, s) = \cosh(10x) + 0.1s \sinh(10x)$$

The partial derivative $\frac{\partial y}{\partial s} = 0.1 \sinh(10x)$ can be as large as $0.1 \sinh(100) \approx 1.3 \cdot 10^{42}$ on the interval $[0, 10]$. The slope for which the boundary condition at $x = 10$ is satisfied is

$$s = \frac{10(B - \cosh(100))}{\sinh(100)}$$

Substituting this value of $s$ into the general solution of the IVP to find the solution $y(x)$ of the BVP, we then find that

$$\left| \frac{\partial y}{\partial B} \right| = \left| \frac{\sinh(10x)}{\sinh(100)} \right| \leq 1$$

Evidently the solution of the IVP is considerably more sensitive to changes in the initial slope $y'(0) = s$ than the solution of the BVP is to changes in the boundary value $y(10) = B$. If the IVPs are not too unstable, shooting can be quite effective. Unstable IVPs can cause a shooting code to fail because the integration "blows up" before reaching the end of the interval. More often, though, the IVP solver reaches the end of the interval but is unable to compute an accurate result there. Because the nonlinear algebraic equations cannot be evaluated accurately, the nonlinear equation solver cannot find accurately the initial values needed to determine the solution of the BVP. By proper preparation of singular points and infinite ranges as illustrated in this chapter, simple shooting codes can be applied to many more problems than is generally appreciated. Nevertheless, they are limited to problems for which the IVPs are moderately stable, and as a consequence there are very few in wide use. The NAG library code D02SAF (Gladwell 1987) and its drivers are simple shooting codes that can be useful because they deal so conveniently with an exceptionally large class of BVPs. Example 3.5.3 illustrates this.

One way to overcome difficulties due to instability of the IVPs that arise when shooting is to break the range of integration into several parts. To appreciate the potential of this technique, recall that the bound we considered in Chapter 2 for the stability of an IVP with a function $f(t, y)$ that satisfies a Lipschitz condition with constant $L$ involves a factor $e^{L(b-a)}$. As least as far as this *bound* is concerned, reducing the length of the interval of integration gives an exponential improvement in the stability of the IVP. After breaking up the interval of integration, the ODEs are integrated independently over the various pieces. The solutions on adjacent subintervals are then required to have the same value at the break point they have in common. With this, the approximate solutions on the various pieces together form a continuous approximation on the whole interval. The boundary conditions and the continuity conditions form a set of nonlinear algebraic equations with $(N + 1)d$ unknowns. Here $d$ is the number of first-order ODEs and $N$ is the number of break points. This approach is called *multiple shooting*. Much as when evaluating implicit methods for IVPs, the algebraic equations are solved with a variant of Newton's method.

At each iteration, a highly structured system of linear equations is solved. It is important to recognize and exploit the structure because the system can be large. If the IVPs are unstable enough to warrant multiple shooting, then some form of pivoting is needed to solve the linear equations stably. A considerable amount of research has been devoted to the stable solution of large linear systems with the structure corresponding to multiple shooting. If the break points are chosen carefully and the linear systems are handled properly, multiple shooting can be quite effective. A major challenge in developing a multiple shooting code is to develop algorithms for choosing an initial set of break points, moving them, and adding or deleting them. Examples of software for nonlinear BVPs based on multiple shooting are MUSN by Mattheij & Staarink (1984a,b) and DD04 by England & Reid (*H2KL*).

In this view of multiple shooting, the IVPs on each subinterval are solved with a quality IVP code. Because the IVP solver varies the step size, it is not known in advance how many mesh points the solver will use in each subinterval nor where it will place them. Finite difference methods have a number of different forms. A very popular one can be viewed as multiple shooting with a one-step method taken to the extreme of just one step between break points. More specifically, these finite difference methods approximate the ODEs (3.1) with a solution $y_0, y_1, \ldots, y_N$ computed with a one-step method on each subinterval $[x_i, x_{i+1}]$ of a mesh $a = x_0 < x_1 < \cdots < x_N = b$. The boundary conditions (3.2) become $g(y_0, y_N) = 0$. The trapezoidal rule is a simple and important example that approximates the ODEs by

$$y_{i+1} - y_i = \frac{h_i}{2}[f(x_i, y_i) + f(x_{i+1}, y_{i+1})]$$

for $i = 0, 1, \ldots, N - 1$. These $N$ equations, together with the one for the boundary conditions, constitute a system of nonlinear algebraic equations for the values $y_i \approx y(x_i)$. In this approach, the solution is defined implicitly even when an explicit one-step IVP method like the (forward) Euler method

$$y_{i+1} - y_i = h_i f(x_i, y_i)$$

is used. This is the result of a fundamental difference between IVPs and BVPs. All the information that specifies the solution of an IVP is given at a single point, and the solution evolves from that point in a direction of interest. The information is given at two or more points for BVPs, so there is no "direction" and it is necessary to account for the information at all points when computing the solution. Because explicit one-step IVP formulas are not explicit when used to solve BVPs, they are not nearly as attractive as they are for solving IVPs. When solving BVPs we are not so concerned about the practical difficulties of evaluating implicit one-step methods, so we can take advantage of their good accuracy and stability. In particular, the trapezoidal rule is a second-order formula with good stability that treats both directions alike.

As with multiple shooting, when using the trapezoidal rule there are $(N + 1)d$ non-linear equations, but $N$ is now (very) much larger and we must pay close attention to the form of the equations if the method is to be practical. In practice, the nonlinear equations are solved by a variant of Newton's method. The structure of the linear systems that must be solved at each iteration of Newton's method is more obvious if we assume that both the equations and the boundary conditions are linear, namely,

$$y' = J(x)y + q(x) \tag{3.11}$$

$$B_a y(a) + B_b y(b) = \beta \tag{3.12}$$

The trapezoidal rule on $[x_i, x_{i+1}]$ is then

$$\left[ -I - \frac{h_i}{2} J(x_i) \right] y_i + \left[ I - \frac{h_i}{2} J(x_{i+1}) \right] y_{i+1} = \frac{h_i}{2} [q(x_i) + q(x_{i+1})] \tag{3.13}$$

After scaling the equations, they can be written in matrix form as

$$
\begin{pmatrix}
S_0 & R_0 & & & \\
& S_1 & R_1 & & \\
& & \ddots & \ddots & \\
& & & S_{N-1} & R_{N-1} \\
B_a & & & & B_b
\end{pmatrix}
\begin{pmatrix}
y_0 \\ y_1 \\ \vdots \\ y_{N-1} \\ y_N
\end{pmatrix}
=
\begin{pmatrix}
v_0 \\ v_1 \\ \vdots \\ v_{N-1} \\ \beta
\end{pmatrix}
\tag{3.14}
$$

Here

$$S_i = -\frac{2}{h_i} I - J(x_i), \quad R_i = \frac{2}{h_i} I - J(x_{i+1}), \quad v_i = q(x_i) + q(x_{i+1})$$

for $i = 0, 1, \ldots, N - 1$. One way to exploit the structure of this matrix is to treat it as a general sparse matrix. This is easy in MATLAB, and that is what its solver bvp4c does. Most solvers proceed differently. They solve only problems with separated boundary conditions,

$$B_a y(a) = \beta_a, \quad B_b y(b) = \beta_b$$

because it is easier to deal with the structure of the resulting linear systems.

When the boundary conditions are separated, the equations can be written as

$$
\begin{pmatrix}
B_a & & & & \\
S_0 & R_0 & & & \\
& S_1 & R_1 & & \\
& & \ddots & \ddots & \\
& & & S_{N-1} & R_{N-1} \\
& & & & B_b
\end{pmatrix}
\begin{pmatrix}
y_0 \\ y_1 \\ y_2 \\ \vdots \\ y_{N-1} \\ y_N
\end{pmatrix}
=
\begin{pmatrix}
\beta_a \\ v_0 \\ v_1 \\ \vdots \\ v_{N-1} \\ \beta_b
\end{pmatrix}
$$

It is easy and reasonably efficient to store this matrix as a banded matrix. Gaussian elimination with row pivoting is an effective way to solve banded linear systems. More elaborate and efficient schemes for handling such matrices take account of their *almost–block diagonal* structure and use both column and row pivoting to minimize the storage and preserve the stability of the computation. A survey of these schemes is found in Amodio et al. (2000). The structure is less regular than it might appear because the submatrices $B_a$ and $B_b$ do not have $d$ rows (they are rectangular). That is because, when the boundary conditions are separated, the number of rows in the submatrix $B_a$ corresponds to the number of boundary conditions specified at the end point $a$ and similarly for $B_b$; the two submatrices have a total of $d$ nonzero rows corresponding to the boundary conditions. One reason these more restricted structures are preferred in popular solvers is that the necessary storage for the stable solution of the corresponding linear systems can be determined in advance, which is not the case for the stable solution of general sparse systems.

As a one-step method, the trapezoidal rule is of order 2, meaning that the solution $y(x)$ satisfies the formula (3.13) with a local truncation error $\tau_i$ at $x_i$ that is $O(h_i^3)$. If we define the error $e_i = y(x_i) - y_i$ and subtract the equations (3.14) satisfied by the $y_i$ from the equations satisfied by the $y(x_i)$, we find that

$$\begin{pmatrix} S_0 & R_0 & & & \\ & S_1 & R_1 & & \\ & & \ddots & \ddots & \\ & & & S_{N-1} & R_{N-1} \\ B_a & & & & B_b \end{pmatrix} \begin{pmatrix} e_0 \\ e_1 \\ \vdots \\ e_{N-1} \\ e_N \end{pmatrix} = \begin{pmatrix} \sigma_0 \\ \sigma_1 \\ \vdots \\ \sigma_{N-1} \\ 0 \end{pmatrix} \tag{3.15}$$

Here the vectors $\sigma_i$ are the scaled truncation errors $(2/h_i)\tau_i$ and there is no truncation error in the boundary conditions. From this we see that the vector of errors at the mesh points is equal to the inverse of the matrix times the vector of scaled truncation errors. The finite difference method is said to be *stable* when there is a *uniform* bound on a norm of this inverse matrix. For $h = \max_i h_i$, the scaled truncation errors are $O(h^2)$ and we conclude that the finite difference method based on the trapezoidal rule is convergent of second order if it is stable. The hard part of showing convergence is proving that the finite difference method is stable. It can be shown that if the BVP consisting of the ODE (3.11) and boundary conditions (3.12) has a unique solution and if the ODEs are sufficiently smooth, then the stability of the finite difference scheme follows from the stability of the one-step method for IVPs. Further, the finite difference scheme converges for BVPs at the same rate that the one-step method does for IVPs. The proof of this general result is rather technical and it is the result that concerns us here rather than the proof, so we refer you to more advanced texts like Ascher et al. (1995) and Keller (1992) for the details. Similarly, we have discussed only BVPs with linear equations and linear boundary conditions. They show what is going on and again we refer to more advanced texts to see how the proofs are modified to deal with nonlinear problems.

Most problems are solved more efficiently with methods of order higher than the second-order trapezoidal rule. In a step from $x_i$ of size $h_i$, a Runge–Kutta method of $s$ stages forms $s$ intermediate values $y_{i,j} \approx y(x_{i,j})$ at the points $x_{i,j} = x_i + \alpha_j h_i$. In Chapter 2 we were most interested in explicit RK methods for which these stages can be computed successively, but in general they are determined simultaneously by solving a system of $s$ nonlinear algebraic equations

$$y_{i,j} = y_i + h_i \sum_{k=1}^{s} \beta_{j,k} f(x_{i,k}, y_{i,k}) \tag{3.16}$$

The new approximate solution is then

$$y_{i+1} = y_i + h_i \sum_{j=1}^{s} \gamma_k f(x_{i,j}, y_{i,j})$$

Implicit Runge–Kutta (IRK) formulas that reduce to Gaussian quadrature rules when solving the quadrature problem $y' = f(x)$ are popular in this context because they have excellent stability and achieve the highest possible order for a method with $s$ stages, namely $2s$. In discussing some of the simpler formulas, it is convenient to denote the midpoint of $[x_i, x_{i+1}]$ by $x_{i+1/2}$ and an intermediate approximation there by $y_{i+1/2}$. With this notation, the lowest-order example of the formulas of Gaussian type is the midpoint rule,

$$y_{i+1} - y_i = h_i f(x_{i+1/2}, y_{i+1/2})$$

which has order 2. The Gaussian formulas do not evaluate at the ends of the subinterval. It was pointed out in Section 3.3.1 that this can be helpful when solving a problem with a singularity at an end point.

Another attractive family of IRK formulas reduce to Lobatto quadrature rules when applied to $y' = f(x)$. The lowest-order example is the trapezoidal rule. The next higher-order Lobatto formula is called the Simpson formula because it reduces to the Simpson quadrature rule. The general form displayed in Table 3.1 can be stated in a way that shows more clearly what must be computed by noting that the first stage is equal to $y_i$ and the last to $y_{i+1}$. Then, with the notation introduced for the midpoint, the formula is

$$y_{i+1/2} = y_i + h_i \left[ \frac{5}{24} f(x_i, y_i) + \frac{1}{3} f(x_{i+1/2}, y_{i+1/2}) - \frac{1}{24} f(x_{i+1}, y_{i+1}) \right]$$

$$y_{i+1} = y_i + h_i \left[ \frac{1}{6} f(x_i, y_i) + \frac{2}{3} f(x_{i+1/2}, y_{i+1/2}) + \frac{1}{6} f(x_{i+1}, y_{i+1}) \right]$$

Broadly speaking, all that we have done for the trapezoidal rule applies also to finite difference methods based on a Runge–Kutta formula. There are, however, important issues

Table 3.1: *The Simpson formula of order 4.*

| 0 | 0 | 0 | 0 |
|---|---|---|---|
| $\frac{1}{2}$ | $\frac{5}{24}$ | $\frac{1}{3}$ | $-\frac{1}{24}$ |
| 1 | $\frac{1}{6}$ | $\frac{2}{3}$ | $\frac{1}{6}$ |
| | $\frac{1}{6}$ | $\frac{2}{3}$ | $\frac{1}{6}$ |

that arise in implementing such formulas. An approach to the efficient implementation of formulas of high order called *deferred correction* evaluates them as corrections to a formula that is more easily evaluated. The first robust software for BVPs based on deferred correction was the PASVA3 code of Lentini & Pereyra (1974). The basic formula of this code is the trapezoidal rule. The higher-order formulas are evaluated by successively approximating terms in an expansion of the truncation error of the trapezoidal rule. Cash and his colleagues have developed a number of effective solvers – for example, TWPBVP (Cash & Wright 1991) – based on deferred correction that match carefully a basic (implicit) Runge–Kutta formula and formulas of higher order. The basic formula of TWPBVP is Simpson's rule.

Other popular solvers work directly with IRK methods of order higher than 2. A key issue is the computation of the intermediate approximations $y_{i,j}$. From their very definition, it is clear that we must be able to solve for them in terms of $y_i$ and $y_{i+1}$. Eliminating the intermediate approximations is called *condensation*. This is important in practice because it reduces significantly the size of the global system. In the case of Simpson's formula, condensation can be performed analytically:

$$y_{i+1} = y_i + \frac{h_i}{6}\left[ f(x_i, y_i) + f(x_{i+1}, y_{i+1}) \right.$$
$$\left. + 4f\left(x_{i+1/2}, \frac{y_{i+1} + y_i}{2} + \frac{h_i}{8}[f(x_i, y_i) - f(x_{i+1}, y_{i+1})]\right)\right] \qquad (3.17)$$

Simpson's formula is implemented with analytical condensation in bvp4c. In general, condensation is performed numerically. Substitution of the linear ODE function $J(x)y + g(x)$ for $f(x, y)$ in (3.16) shows what happens. There is a system of linear equations for the intermediate stages $y_{i,j}$ that is coupled only to $y_i$ and $y_{i+1}$. This subsystem can be solved to eliminate these unknowns in the full system. When applied to first-order systems, the solvers COLSYS (Ascher, Christiansen, & Russell 1979, 1981) and COLNEW (Ascher et al. 1995; Bader & Ascher 1987) can be viewed as implementing the family of

IRK methods of Gaussian type. They eliminate the intermediate values numerically when solving for values at the mesh points.

Finite difference methods provide solutions only at mesh points. The issue of obtaining approximate solutions at other points arose earlier when we were solving IVPs with explicit Runge–Kutta methods. There we learned that, for some Runge–Kutta formulas, continuous extensions have been developed that provide an accurate solution throughout a step. A natural continuous extension for an IRK method of $s$ stages defines a polynomial $S(x)$ by the interpolation conditions $S(x_i) = y_i$ and $S'(x_{i,j}) = f(x_{i,j}, y_{i,j})$ for $j = 1, 2, \ldots, s$. For a class of IRK methods that includes those based on quadrature formulas of Gaussian type, it is not difficult to show that this polynomial satisfies $S(x_{i+1}) = y_{i+1}$. This implies that the natural continuous extension is *continuous*; that is, $S(x) \in C[a, b]$. It can also be shown that $S(x_{i,j}) = y_{i,j}$. Along with the interpolation condition, this tells us that $S(x)$ satisfies the ODEs at each intermediate point: $S'(x_{i,j}) = f(x_{i,j}, S(x_{i,j}))$. This property is called *collocation*. We did not pursue these natural continuous extensions in Chapter 2 because generally they do not have the same order of accuracy as the formula. Indeed, for the formulas of Gaussian type, they have only half the order of the basic formula. The midpoint rule provides a simple example. On the subinterval $[x_i, x_{i+1}]$, the continuous extension $S(x)$ is a linear polynomial with $S(x_i) = y_i$ and $S'(x_{i+1/2}) = f(x_{i+1/2}, y_{i+1/2})$. The definition of the midpoint rule tells us that

$$f(x_{i+1/2}, y_{i+1/2}) = \frac{y_{i+1} - y_i}{h_i}$$

The linear polynomial $S(x)$ has a constant slope, so it is clear from this expression for the slope at the midpoint that $S(x)$ is the straight line that interpolates the numerical solution at both ends of the interval. This continuous extension $S(x)$ is $C[a, b]$, but it is clearly not $C^1[a, b]$. It is only first-order accurate between mesh points.

Generally the natural continuous extension is only $C[a, b]$, but the formulas of Lobatto type collocate at both ends of each subinterval; that is,

$$S'(x_i) = f(x_i, S(x_i)) = f(x_i, y_i), \quad S'(x_{i+1}) = f(x_{i+1}, S(x_{i+1})) = f(x_{i+1}, y_{i+1})$$

It follows from this that $S(x) \in C^1[a, b]$ for these formulas. As it turns out, the natural continuous extension of the Simpson formula preserves the order of the formula itself. The Simpson formula is an attractive method for solving BVPs in MATLAB because a numerical solution that is $C^1[a, b]$ and uniformly fourth-order accurate is well suited to the graphical study of solutions. The scheme has other attractions. One is that analytical condensation is an efficient way to evaluate the formula in MATLAB.

A classic approach to solving BVPs is to choose a form for the approximate solution $S(x)$ that involves parameters determined by requiring that $S(x)$ satisfy the boundary conditions and then to collocate the ODEs at sufficiently many points. We have seen that some

important classes of finite difference formulas can be viewed as resulting from collocation with a continuous piecewise-polynomial function $S(x)$ (a spline). That is the view
taken in popular solvers; in particular, that is why bvp4c is described as a collocation
code. In this view, bvp4c solves BVPs of the form (3.1) and (3.2) by computing a continuous function $S(x)$ that is a cubic polynomial on each subinterval $[x_i, x_{i+1}]$ of a mesh
$a = x_0 < x_1 < \cdots < x_N = b$. The coefficients of the cubic polynomials that make up
this function are determined by requiring that $S(x)$ be continuous on $[a, b]$, satisfy the
boundary conditions

$$g(S(a), S(b)) = 0$$

and satisfy the ODEs at both end points and the midpoint of each subinterval,

$$S'(x_i) = f(x_i, S(x_i))$$
$$S'(x_{i+1/2}) = f(x_{i+1/2}, S(x_{i+1/2}))$$
$$S'(x_{i+1}) = f(x_{i+1}, S(x_{i+1}))$$

As pointed out earlier, the collocation conditions at the ends of the subinterval imply that
$S(x) \in C^1[a, b]$. All together these conditions result in a system of nonlinear algebraic
equations for the coefficients of the cubic polynomials that make up $S(x)$. When the details are worked out, it is found that this function $S(x)$ is the natural continuous extension
of the Simpson formula. Accordingly, we can regard this method either as a collocation
method or as a finite difference method with a continuous extension. Enright & Muir
(1996) implement a family of IRK formulas with continuous extensions in the Fortran
code MIRKDC. One member of the family corresponds to the Simpson formula, but its
continuous extension is a polynomial of higher degree and higher accuracy than the natural one used in bvp4c. Clearly the collocation approach is not restricted to systems of
first-order ODEs, and there are some advantages to treating higher-order ODEs directly.
The codes COLSYS and COLNEW cited earlier in connection with finite difference methods are distinctive because they treat higher-order ODEs directly. It is only when they are
applied to first-order systems of ODEs that their methods can be viewed as equivalent to
a finite difference scheme based on IRK methods of Gaussian type.

   Because a BVP can have more than one solution, it is necessary to supply codes with a
guess that identifies the solution of interest. In all the popular codes, the numerical solution of a nonlinear BVP is accomplished directly or indirectly by linearization. The codes
use devices to enhance the rate of convergence, but a good guess may be necessary to obtain convergence. The guess involves supplying both a mesh that reveals the behavior of
the solution and either values of the guessed solution on the mesh or a function for computing them. After obtaining convergence for this mesh, the codes adapt the mesh so as
to obtain an accurate numerical solution with a modest number of mesh points. In some
respects this is much more difficult than solving IVPs. For IVPs the most difficult part of

step-size adjustment is getting on scale at the first step, because the steps that follow are adjusted one at a time and only slow variation is permitted. For BVPs the most difficult part is providing an initial approximation to a solution. In large measure this burden is placed on the user, who must provide guesses for the mesh and solution that will lead to convergence. Let us now discuss briefly how the solvers estimate and control the error.

A natural approach to error control is to estimate the truncation error and adjust the mesh accordingly. When the truncation error for the subinterval $[x_i, x_{i+1}]$ can be expressed in terms of a derivative of the solution, this derivative can be approximated by interpolating the values $y_i$ and $y_{i+1}$ (as well as some approximations from neighboring intervals) and differentiating the interpolant. Recall that we did something like this to approximate the truncation error of the first-order BDF formula in Chapter 2. An important distinction when solving BVPs is that we can use values on both sides of the subinterval. The truncation error of the trapezoidal rule has this form, and Lentini and Pereyra estimate it in this way in PASVA3. Indeed, the same is done for the higher-order formulas of this deferred correction code. The truncation errors of the Gaussian IRK formulas also have this form, and this way of estimating the errors is used in the codes COLSYS and COLNEW. Another way of estimating truncation error that we studied in Chapter 2 is to compare the result of one formula to the result of a higher-order formula. This is quite natural in the way that Cash and Wright use deferred correction in the code TWPBVP, because each formula is evaluated as a correction to a lower-order result.

With estimates for the truncation errors, the step sizes can be adjusted much as when solving IVPs. An obvious difference is that when solving BVPs the entire mesh is changed whereas only one step at a time is changed when solving IVPs. Because BVPs are global in nature, refining the mesh in a region where the truncation errors are large affects the numerical solution throughout the interval. It is possible that the numerical solution is improved in one region but worsened elsewhere. For methods with a truncation error on $[x_i, x_{i+1}]$ that depends on a derivative of the solution in this subinterval, the effects of a local change in a mesh are local (at least to leading order), and thus changing the mesh to reduce the truncation errors in a region actually improves the overall solution.

A serious difficulty with BVP solvers is that the schemes for approximating truncation errors and adjusting the mesh depend on the mesh being sufficiently fine. It is important, then, that the solver refine the mesh in a way that is plausible even when the estimates of the truncation errors are very poor, as they often will be when the mesh is too crude or not well adapted to the behavior of the solution. To improve reliability, the codes COLSYS and COLNEW supplement control of truncation error with another assessment of the error. After it appears that a solution has been found to an appropriate accuracy of the truncation error, the mesh is halved and another solution computed. The error of this second solution is estimated by a process called *extrapolation,* which proceeds as follows. Suppose we can compute a function $F(x; h)$ involving a parameter $h$ and that we want $F(x; 0)$, the limit as $h \to 0$. If we know that the error $e(x; h) = F(x; h) - F(x; 0)$ behaves like $\phi(x)h^p$ as

$h \rightarrow 0$, then we can use approximations $F(x; h)$ and $F(x; h/2)$ computed for specific $h$ and $h/2$ in order to estimate the error of the more accurate result. Our assumption about how the error depends on $h$ tells us that

$$e\left(x; \frac{h}{2}\right) \sim \phi(x)h^p 2^{-p}$$

and hence

$$F(x; h) - F\left(x; \frac{h}{2}\right) = e(x; h) - e\left(x; \frac{h}{2}\right) \sim \phi(x)h^p[1 - 2^{-p}]$$

Solving this relation for $\phi(x)h^p$ provides a computable estimate of the error of the more accurate result,

$$e\left(x; \frac{h}{2}\right) \sim \frac{1}{2^p - 1}\left[F(x; h) - F\left(x; \frac{h}{2}\right)\right] \tag{3.18}$$

In applying this principle to BVPs, $h$ is the maximum step size and $p$ is the order of the method. If the function $f(x, y)$ defining the ODE is sufficiently smooth and if the mesh is fine enough that the leading term dominates in an expansion of the error, then extrapolation furnishes a way to estimate the error and so to confirm that the numerical solution has the desired accuracy.

The MIRKDC solver of Enright and Muir and the bvp4c solver of MATLAB take an unusual approach to the control of error that is intended to deal more robustly with poor guesses for the mesh and solution. They produce approximate solutions $S(x) \in C^1[a, b]$. For such an approximation, the residual in the ODEs is defined by

$$r(x) = S'(x) - f(x, S(x))$$

Similarly, the residual in the boundary conditions is $\delta = g(S(a), S(b))$. Put differently, the approximation $S(x)$ is an *exact* solution of the BVP

$$Y' = f(x, Y) + r(x), \quad g(Y(a), Y(b)) - \delta = 0$$

From the point of view of *backward error analysis,* $S(x)$ is an accurate solution if it is an exact solution of a BVP that is "close" to the one given, meaning that the perturbations $r(x)$ and $\delta$ are "small". If the BVP is reasonably well-conditioned then small changes to the problem must result in small changes to the solution, so a solution that is accurate in the sense of backward error analysis is also accurate in the usual sense of the approximate solution being close to the true solution. Both MIRKDC and bvp4c control the sizes of these residuals. The approach is attractive because the residuals for these methods are well-defined no matter how crude the mesh. Further, the residual $r(x)$ can be evaluated at any point $x$ that we wish, so we can approximate its size as accurately as we wish (and are willing to pay for). For robustness, bvp4c measures the size of the residual on each subinterval $[x_i, x_{i+1}]$ by an integral

$$\int_{x_i}^{x_{i+1}} \|r(x)\|^2 \, dx$$

This integral is approximated with a five-point Lobatto quadrature formula. Using the fact that the residual vanishes at both end points and the midpoint of the interval, this quadrature formula requires only two additional evaluations of $r(x)$ and hence of $f(x, S(x))$. It gives a plausible estimate of the size of the residual for any mesh, and it is asymptotically correct as $h_i \to 0$. One of the useful properties of the Simpson formula with its natural continuous extension is that the effects on the residual of local changes to the mesh are local, at least to leading order. Proofs of these technical matters can be found in Kierzenka (1998) and Kierzenka & Shampine (2001).

In passing, we remark that a simple shooting code can also be described as controlling residuals. At each step the IVP solver controls the local error, which is equivalent to controlling the size of the residual of an appropriate continuous extension of its numerical method. The nonlinear algebraic equations of shooting state that initial values are to be found for which the boundary conditions are satisfied. Thus, the nonlinear equation solver finds initial values for which the numerical solution has a small residual in the boundary conditions. The IVP solver and the nonlinear equation solver of a shooting code produce, then, a numerical solution that satisfies the boundary conditions and the ODEs with small residuals. Something similar can be said of multiple shooting.

### ■ EXERCISE 3.10
Show that the BVP
$$y'' + 100y = 0$$

with boundary conditions $y(0) = 1$ and $y(10) = B$ is insensitive to small changes in the value of $B$.

### ■ EXERCISE 3.11
By expanding in Taylor series, find an expression of the form $\sigma_i = K h^p y^{(p+1)}(\eta_i)$ for the scaled truncation errors in equation (3.15). That is, find the integer $p$ and the constant $K$.

### ■ EXERCISE 3.12
Verify the expression for the condensed version of Simpson's formula given in equation (3.17).

### ■ EXERCISE 3.13
Write out the system of linear algebraic equations that arise when using the condensed Simpson's formula (3.17) for solving the BVP consisting of the linear ODE (3.11) with boundary conditions (3.12).

■  **EXERCISE 3.14**

Show that the function $S(x)$ associated with the BVP solver `bvp4c` (as defined on page 164) is continuously differentiable on the interval $[a, b]$. That is, show that $S(x) \in C^1[a, b]$.

# 3.5  Solving BVPs in MATLAB

In this section we use a variety of examples from the literature to illustrate the numerical solution of BVPs with the MATLAB solver `bvp4c`. BVPs arise in such diverse forms that many require some preparation for their solution by standard software – and some require extensive preparation. Several examples illustrate how to do this for the most common forms and the most widely used BVP solvers. Other examples show how you might be able to speed up significantly the solution of a BVP with `bvp4c`. Exercises are provided both for instruction and practice.

## EXAMPLE 3.5.1

Bratu's equation arises in a model of spontaneous combustion and is mathematically interesting as an example of bifurcation simple enough to solve in a semianalytical way (Davis 1962). The differential equation is

$$y'' + \lambda e^y = 0$$

with boundary conditions $y(0) = 0 = y(1)$. Bratu showed that, for values of $\lambda$ such that $0 \leq \lambda < \lambda^* = 3.51383\ldots$, there are two solutions. Both solutions have a parabolic form and are concave down. These solutions grow closer as $\lambda \to \lambda^*$ and coalesce to give a unique solution when $\lambda = \lambda^*$. There is no solution when $\lambda > \lambda^*$. To show how to use `bvp4c`, we solve this BVP when $\lambda = 1$. This is much like solving an IVP, so we state a program and then discuss those matters that are different.

```
function sol = ch3ex1

solinit = bvpinit(linspace(0,1,5),@guess);
sol = bvp4c(@odes,@bcs,solinit);
plot(sol.x,sol.y(1,:));
figure
xint = linspace(0,1,100);
Sxint = deval(sol,xint);
plot(xint,Sxint(1,:));

%==================================================
```

```
function v = guess(x)
v = [ x*(1-x); 1-2*x ];

function dydx = odes(x,y)
dydx = [ y(2); -exp(y(1)) ];

function res = bcs(ya,yb)
res = [ ya(1); yb(1) ];
```

In simplest use bvp4c has only three arguments: a function odes for evaluating the ODEs, a function bcs for evaluating the residual in the boundary conditions, and a structure solinit that provides a guess for a mesh and the solution on this mesh. The ODEs are handled exactly as in the MATLAB IVP solvers and so require no further discussion here. Boundary conditions $g(y(a), y(b)) = 0$ are coded so that, when the function bcs is given the current approximations $ya \approx y(a)$ and $yb \approx y(b)$, it evaluates and returns the residual $g(ya, yb)$ as a column vector.

When solving BVPs you must provide a guess for the solution, both to identify the solution that interests you and to assist the solver in computing that solution. The guess is supplied to bvp4c as a structure formed by the auxiliary function bvpinit. The first argument of bvpinit is a guess for a mesh that reveals the behavior of the solution. Here we try five equally spaced points in $[0, 1]$. The second argument is a guess for the solution on the specified mesh. Here the solution has two components, $y(x)$ and $y'(x)$. This guess can be provided in two ways. Here we see them provided by means of a function guess. The guess $x(1 - x)$ for $y(x)$ has the correct shape and satisfies the boundary conditions. The derivative of this guess for $y(x)$ is used as the guess for $y'(x)$. Alternatively, if we take all the estimated solution components to be constant then we can provide the vector of these constants directly. This is convenient and often works. Here, for example, you can solve the BVP by providing the vector [0.5; 0] to bvpinit instead of the function guess.

The code bvp4c returns a solution structure, here called sol. This is an option with the IVP solvers, but it is the only form of output from bvp4c. As with the IVP solvers, the field sol.x contains the mesh found by the solver and the field sol.y contains the solution on this mesh. Figure 3.2 shows the result of plotting the data returned in these fields. The cost of solving a BVP depends strongly on the number of mesh points, so bvp4c uses no more than necessary. This BVP for Bratu's equation is so easy that the solver computes an accurate solution using only the five mesh points of the guess. The values plotted are accurate, so we need only approximate the solution at more points in order to plot a smooth graph. The solver bvp4c produces a solution $S(x) \in C^1[0, 1]$ and returns in sol all the information needed to evaluate this smooth approximate solution. Just as with the IVP solvers, this is done with the auxiliary function deval. Recall that the first argument of deval is the solution structure and the second is an array of points where the solution

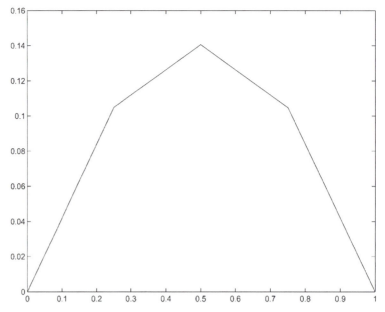

Figure 3.2: Solution of Bratu's problem plotted at mesh points only.

is desired. Here we evaluate the approximate solution $S(x)$ at 100 equally spaced points and obtain the smooth graph in Figure 3.3. The function `deval` is evaluating a piecewise-cubic function and it is vectorized, so obtaining even a great many solution values for plotting is inexpensive.

This BVP for Bratu's equation has two solutions. If you multiply the guess for the solution and its derivative coded in `ch3ex1.m` by a factor of 5, you will again compute the solution displayed. If you multiply them both by 20, you will compute the other solution. It is rather larger than the solution plotted, having a maximum of about 4.1. If you multiply the guess for the solution and its derivative by a factor of 100, the solver will fail to converge. This illustrates the fact that which solution you compute – or even whether you compute a solution at all – depends strongly on the initial guess. In Exercise 3.15 you are asked to explore a conservation law satisfied by the solution of this problem. The cannon problem of Exercise 3.16 is much like this BVP.

## EXAMPLE 3.5.2

A nonlinear eigenproblem of lubrication theory that is solved in Keller (1992, sec. 6.1) involves only a single first-order ODE,

$$\varepsilon y' = \sin^2(x) - \lambda \frac{\sin^4(x)}{y} \tag{3.19}$$

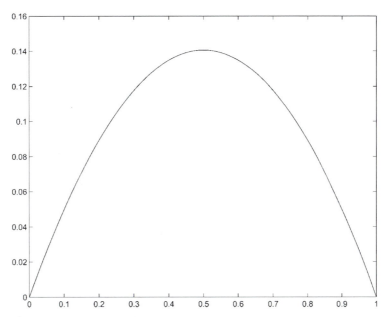

Figure 3.3: Solution of Bratu's problem plotted at 100 equally spaced points.

Here $\varepsilon$ is a known parameter, and we compute a value of the unknown parameter $\lambda$ for which there is a solution of the ODE that satisfies the two boundary conditions

$$y\left(-\frac{\pi}{2}\right) = 1, \quad y\left(\frac{\pi}{2}\right) = 1 \qquad (3.20)$$

These boundary conditions may be viewed as two equations, one for the first-order ODE and one for the unknown parameter. Unknown parameters are not at all unusual. Sometimes they arise in the physical model; often they are introduced to facilitate computing a solution in difficult circumstances. Although bvp4c makes it easy to solve BVPs involving unknown parameters, most solvers do not, so after illustrating the use of bvp4c we'll explain how to use solvers that do not provide for unknown parameters.

Just as with the solution, when there are unknown parameters you must supply estimates for them – both to identify the particular set of parameters that interests you and to assist the solver in computing them. The estimates are supplied in a vector as the third argument of bvpinit. Correspondingly, you must include the vector of unknown parameters as the third argument to the functions for evaluating the ODEs and the residual in the boundary conditions. You must do this for each function even if it makes no use of the unknown parameters. When there are unknown parameters, the solution structure has a field named parameters that contains the vector of parameters computed by the solver. The program ch3ex2.m solves the BVP with ODE defined by (3.19) and boundary conditions defined by (3.20) for the value $\varepsilon = 0.1$. We guess that $\lambda$ is about 1 and use

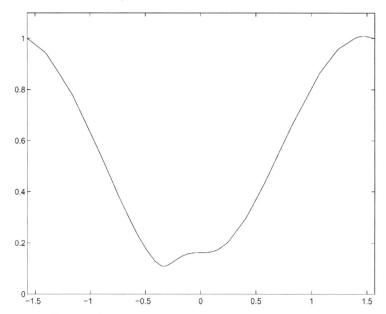

Figure 3.4: Solution of the lubrication problem with $\varepsilon = 0.1$.

a convenient constant estimate of $y(x) \approx 0.5$. Notice that `lambda` is an input argument of `bcs` even though it is not used by this function. The program reports to the screen that $\lambda \approx 1.01864$ and produces the graph of Figure 3.4.

```
function sol = ch3ex2

solinit = bvpinit(linspace(-pi/2,pi/2,20),0.5,1);
sol = bvp4c(@ode,@bcs,solinit);
fprintf('lambda = %g.\n',sol.parameters)
plot(sol.x,sol.y(1,:));
axis([-pi/2 pi/2 0 1.1]);


%=================================================
function dydx = ode(x,y,lambda)
epsilon = 0.1;
dydx = (sin(x)^2 - lambda*sin(x)^4/y)/epsilon;

function res = bcs(ya,yb,lambda)
res = [ ya-1; yb-1 ];
```

Most solvers do not provide for unknown parameters, but it is easy to prepare a problem that has parameters so that these solvers can be used. For the problem at hand, we have an unknown function $y_1(x) = y(x)$ and an unknown parameter $\lambda$. The trick is to introduce a second unknown function, $y_2(x) = \lambda$, and then add a trivial ODE stating that this new function is constant. This trick results in the ODE system

$$y_1' = \varepsilon^{-1}\left[\sin^2(x) - y_2\frac{\sin^4(x)}{y_1}\right]$$

$$y_2' = 0$$

which, along with the two boundary conditions (3.20), is now a BVP without unknown parameters. Many solvers require you to provide analytical partial derivatives for the ODEs and boundary conditions with respect to the solution. For such solvers, introducing new variables for unknown parameters is not the end of the preparation; you must also provide the analytical partial derivatives of both the ODEs and the boundary conditions with respect to the new variables (i.e., with respect to the parameters).

## EXAMPLE 3.5.3

In Seydel (1988) the propagation of nerve impulses is described by the system of ODEs

$$y_1' = 3\left(y_1 + y_2 - \frac{y_1^3}{3} - 1.3\right)$$

$$y_2' = -\frac{1}{3}(y_1 - 0.7 + 0.8y_2)$$

subject to the periodic boundary conditions

$$y_1(0) = y_1(T), \quad y_2(0) = y_2(T)$$

These two nonseparated boundary conditions suffice if we want solutions with a specific period $T$. Exercise 3.30 is an example of this. However, if we do not specify the period then we must find an unknown parameter $T$ along with solutions of period $T$, and for that we need another boundary condition. To better understand this, notice that if $(y_1(t), y_2(t))$ is a periodic solution of these autonomous differential equations then so is $(u_1(t), u_2(t)) = (y_1(t + \gamma), y_2(t + \gamma))$ for any constant $\gamma$. One possible boundary condition is to specify a value for one of the solution components, for example, $y_1(0) = 0$. With this boundary condition we might as well replace the periodicity condition $y_1(0) = y_1(T)$ with $y_1(T) = 0$ because it is equivalent and separated. Thus we solve the ODEs subject to the boundary conditions

$$y_1(0) = 0, \quad y_1(T) = 0, \quad y_2(0) = y_2(T)$$

one of which is nonseparated.

The code `bvp4c` accepts problems with nonseparated boundary conditions, but most solvers require that the boundary conditions be separated. After solving this BVP with bvp4c, we discuss how problems with nonseparated boundary conditions can be prepared for solution with other codes. A complication of this example is that the length of the interval $[0, T]$ is unknown. The shooting code D02SAF (NAG 2002) treats all unknowns – including the end points of the interval, the boundary values, and any parameters – as unknown parameters, so it can solve this problem directly. Since D02SAF is the only solver in wide use with this capability, we must prepare the problem for most solvers (including bvp4c) by transforming it to one formulated on a fixed interval.

If we change the independent variable from $t$ to $x = t/T$, the ODEs become

$$\frac{dy_1}{dx} = 3T\left(y_1 + y_2 - \frac{y_1^3}{3} - 1.3\right)$$

$$\frac{dy_2}{dx} = -\frac{T}{3}(y_1 - 0.7 + 0.8y_2)$$

The new problem is posed on the interval $[0, 1]$, and the new boundary conditions are

$$y_1(0) = 0, \quad y_1(1) = 0, \quad y_2(0) = y_2(1)$$

After this preparation, the BVP is solved easily with the program ch3ex3.m.

```
function sol = ch3ex3

solinit = bvpinit(linspace(0,1,5),@guess,2*pi);
sol = bvp4c(@ode,@bc,solinit);
T = sol.parameters;
fprintf('The computed period T = %g.\n',T);
plot(T*sol.x,sol.y(1,:),T*solinit.x,solinit.y(1,:),'ro')
legend('Computed Solution','Guessed Solution');
axis([0 T -2.2 2]);


%=====================================================
function v = guess(x)
v = [ sin(2*pi*x); cos(2*pi*x)];

function dydt = ode(x,y,T);
dydt = [ 3*T*(y(1) + y(2) - (y(1)^3)/3 - 1.3)
         -(T/3)*(y(1) - 0.7 + 0.8*y(2))          ];

function res = bc(ya,yb,T)
res = [ ya(1); yb(1); (ya(2) - yb(2)) ];
```

We guess that $T = 2\pi$, $y_1(x) \approx \sin(2\pi x)$, and $y_2(x) \approx \cos(2\pi x)$. After computing the solution, the independent variable $x$ is rescaled to the original independent variable

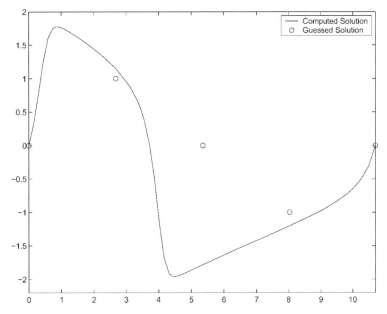

Figure 3.5: Periodic solution of a nerve impulse model.

$t = Tx$ for the plot of Figure 3.5. Plotting of the initial guess shows that it is not very accurate. The program reports to the screen that $T \approx 10.7106$, so $2\pi$ was also not a very good guess for the period $T$. This problem is sufficiently easy that accurate estimates of the solution and the period are not needed for convergence.

The boundary condition $y_2(0) = y_2(T)$ will be used to show how to deal with nonseparated boundary conditions when the BVP solver does not provide for them. The idea is to introduce a new unknown, $y_3(t) \equiv y_2(T)$. This function is constant, so we add $y_3' = 0$ to the system of ODEs. Because $y_3(t)$ is constant, the boundary condition $y_2(0) = y_2(T)$ is equivalent to the boundary condition $y_2(0) = y_3(0)$ and, by definition, we have another boundary condition $y_2(T) = y_3(T)$. In this way a nonseparated boundary condition is replaced by two separated boundary conditions that involve a new solution component, and there is an additional ODE. The final set of ODEs and (separated) boundary conditions are

$$\frac{dy_1}{dx} = 3\left(y_1 + y_2 - \frac{y_1^3}{3} - 1.3\right)$$

$$\frac{dy_2}{dx} = -\frac{1}{3}(y_1 - 0.7 + 0.8y_2)$$

$$\frac{dy_3}{dx} = 0$$

and

$$y_1(0) = 0, \quad y_1(T) = 0, \quad y_2(0) = y_3(0), \quad y_2(T) = y_3(T)$$

In general, each nonseparated boundary condition leads to a new solution component, an additional boundary condition, and a trivial ODE. Exercise 3.30 provides another example.

## EXAMPLE 3.5.4

We use an example of Gladwell (1979a) (with a minor error corrected) to show how unknown parameters can arise when solving BVPs with singularities and how to handle some problems posed on an infinite interval. This example arose in a similarity solution of some partial differential equations describing fluid flow. The analysis is somewhat lengthy, but if you should need to do something similar yourself then you might find the details of this example to be helpful. The system of ODEs is

$$3yy'' = 2(y' - z), \quad z'' = -yz' \tag{3.21}$$

The boundary conditions are

$$y(0) = 0, \quad z(0) = 1 \tag{3.22}$$

and

$$y'(+\infty) = 0, \quad z'(+\infty) = 0 \tag{3.23}$$

The first equation is singular at the origin because $y(0) = 0$. To deal with this, we first work out analytical approximations to the solution components $y(x)$ and $z(x)$ that are accurate on an interval $[0, \delta]$ for some small $\delta > 0$. With two second-order equations and only two boundary conditions at the origin, these approximations will necessarily involve two free parameters, effectively replacing the unknown boundary values $y'(0)$ and $z'(0)$. We then solve a BVP on $[\delta, \infty)$ with the same ODEs (3.21) and same boundary conditions at infinity (3.23). The equations are not singular on this interval. The boundary conditions at the origin (3.22) are replaced in this new problem by the requirement that the numerical solutions at $\delta$ agree with the analytical approximations there. The free parameters in the analytical approximations show up as unknown parameters in the BVP, which we solve numerically. After determining these parameters as part of solving the BVP, we are able to approximate $y(x)$ and $z(x)$ on all of the range $[0, \infty)$.

With a nonlinear differential equation it is not at all obvious how the solutions behave near the singular point. Generally an understanding of the physical problem provides valuable guidance. Without this kind of guidance, we might as well be optimistic and look for analytical approximations in the form of Taylor series. With some computer assistance this is easy enough and provides some insight. Exploiting the MATLAB Symbolic Toolbox functionality, we use the following script to substitute a few terms of Taylor series expansions into the equations (3.21).

```
syms x y z A B C D E F eqn1 eqn2
y = 0 + A*x + B*x^2 + C*x^3;
z = 1 + D*x + E*x^2 + F*x^3;
eqn1 = collect(3*y*diff(y,2,'x') - 2*(diff(y,'x') - z))
eqn2 = collect(diff(z,2,'x') + y*diff(z,'x'))
```

Note that we have already taken into account the boundary conditions (3.22). This script produces the (slightly edited) output

```
eqn1 = 18*C^2*x^4+(24*B*C+2*F)*x^3
       +(-6*C+18*A*C+6*B^2+2*E)*x^2+(-4*B+6*A*B+2*D)*x-2*A+2
eqn2 = 3*C*F*x^5+(3*B*F+2*C*E)*x^4+(3*A*F+2*B*E+C*D)*x^3
       +(2*A*E+B*D)*x^2+(6*F+A*D)*x+2*E
```

We are interested in the behavior as $x \to 0$ and so, the higher the power of $x$, the less effect it has in these expansions. Our goal is to satisfy the equations as well as possible, so we want to choose coefficients that make as many successive terms zero as possible, starting with the lowest power. To eliminate the constant terms, we see from the expansions that we must take $A = 1$ and $E = 0$. Next, to eliminate the terms in $x$, we must take $D = -B$ and $F = -D/6$. We cannot eliminate the terms in $x^2$ without including more terms in the Taylor series. We thus conclude that, for small values of $x$, we have $y(x) = x + Bx^2 + \cdots$ and $z(x) = 1 - Bx + (B/6)x^3 + \cdots$. In these expansions the coefficient $B$ is a free parameter.

This seems to have gone very well, but these series can't be the approximations that we need! After all, the approximations must have two free parameters and these series have only one. It is possible that another free parameter will appear at a higher power, but it is more likely that we were mistaken when we assumed that the solutions can be expanded in Taylor series. Let's try expanding the solutions in powers of $x$ with powers that are not necessarily integers. This is harder, so we begin with just a few terms that we hope will reveal a power that is not an integer:

$$y(x) = ax^\alpha + bx^\beta + \cdots$$
$$z(x) = 1 + cx^\gamma + dx^\delta + \cdots$$

With the necessary assumptions that $0 < \alpha < \beta$ and $0 < \gamma < \delta$, these series satisfy the boundary conditions (3.22). Substituting them into the first differential equation of (3.21) results in the equation

$$0 = 3[ax^\alpha + bx^\beta + \cdots][\alpha(\alpha - 1)ax^{\alpha-2} + \beta(\beta - 1)bx^{\beta-2} + \cdots]$$
$$- 2[(\alpha ax^{\alpha-1} + \beta bx^{\beta-1} + \cdots) - (1 + cx^\gamma + dx^\delta + \cdots)]$$

If $\alpha < 1$, the lowest-order term is $3\alpha(\alpha - 1)a^2 x^{2\alpha-2}$. It cannot be removed because we exclude the possibilities $a = 0$ and $\alpha = 0$. On the other hand, if we suppose that $\alpha = 1$ then there are two terms of lowest order, namely $-2[(ax^0) - (1)]$, that can be removed by taking $a = 1$. Turning now to the second equation, we have

$$0 = \gamma(\gamma - 1)cx^{\gamma-2} + \delta(\delta - 1)dx^{\delta-2} + \cdots$$
$$+ (x + bx^\beta + \cdots)(\gamma cx^{\gamma-1} + \delta dx^{\delta-1} + \cdots)$$

We see that the lowest-order term is $\gamma(\gamma - 1)cx^{\gamma-2}$. To remove it, we must take $\gamma = 1$. With this the equation becomes

$$0 = \delta(\delta - 1)dx^{\delta-2} + \cdots + cx + \cdots$$

To remove the term in $x^{\delta-2}$, we must choose $\delta$ so that the power is the same as that of the "next" term; that is, we must "balance" this term with another. If we take $\delta = 3$, the term is balanced with the term in $x$. We can then let $c$ be a free parameter and take $d = -c/6$ to remove these two terms of lowest order. Let us return now to the expansion of the first equation and substitute all the information gleaned so far:

$$3[x + bx^\beta + \cdots][\beta(\beta - 1)bx^{\beta-2} + \cdots] = 2\left[(\beta bx^{\beta-1} + \cdots) - \left(cx - \tfrac{c}{6}x^\delta + \cdots\right)\right]$$

If $\beta < 2$, the lowest power is $\beta - 1$. Equating the coefficients of $x^{\beta-1}$ to remove these terms, we find that

$$3b\beta(\beta - 1) = 2b\beta$$

hence $\beta = \frac{5}{3}$ and $b$ is a free parameter.

The hard part of deriving the expansions is now over. What we missed in the Taylor series approach is the $bx^{5/3}$ term in the expansion for $y(x)$. Here $b$ is the free parameter that we needed in addition to the free parameter $B$ that we discovered in the Taylor series approach. The crucial matter is to recognize that we must look for series in powers of $x^{1/3}$. With this knowledge it is straightforward to compute more terms. For instance, to compute the next two terms we can use the script

```
syms x y z b c d e eqn1 eqn2
y = x + b*x^(5/3) + d*x^2;
z = 1 + c*x - (c/6)*x^3 + e*x^(10/3);
eqn1 = collect(3*y*diff(y,2,'x') - 2*(diff(y,'x') - z))
eqn2 = collect(diff(z,2,'x') + y*diff(z,'x'))
```

The output, edited to show only the lowest-order terms, is

```
eqn1 = ... +10/3*b^2*x^(4/3)+(2*d+2*c)*x
eqn2 = ... +b*c*x^(5/3)+70/9*e*x^(4/3)
```

To remove the lowest-order terms, we must take $d = -c$ and $e = 0$. Repetition leads to

$$y(x) = x + bx^{5/3} - cx^2 - \frac{5b^2}{7}x^{7/3} + \frac{7bc}{6}x^{8/3} + \left(\frac{95b^3}{126} - \frac{c^2}{2}\right)x^3 + \cdots$$

$$z(x) = 1 + cx - \frac{c}{6}x^3 - \frac{9bc}{88}x^{11/3} + \frac{c^2}{12}x^4 + \frac{9b^2c}{182}x^{13/3} - \frac{3bc^2}{44}x^{14/3} + \cdots$$

It is natural to approximate the boundary conditions at infinity by $y'(\Delta) = 0$ and $z'(\Delta) = 0$ for some large $\Delta$. This kind of approach often works, but if you try it in the program ch3ex4.m you will find that the computation fails. Evidently we need to provide the solver with more information about how solutions behave at infinity. For this we first observe that there is a constant solution of the ODEs that satisfies the boundary conditions at infinity, namely $y(x) \equiv \alpha$ and $z(x) \equiv 0$ for arbitrary $\alpha$. (A solution of this kind with $\alpha = \delta \approx y(\delta)$ is used as the guess in ch3ex4.m.) To study the behavior of solutions of the ODEs that are close to such a solution, we linearize the equations about $(\alpha, 0)$ or, equivalently, look for solutions of the form

$$y(x) \sim \alpha + \beta e^{\gamma x}, \quad z(x) \sim 0 + \delta e^{\varepsilon x}$$

as $x \to \infty$. For $y(x)$ of this form,

$$y'(x) \sim \gamma\beta e^{\gamma x}$$

In order to have a nontrivial solution that satisfies the boundary condition $y'(\infty) = 0$, we must have $\gamma < 0$. Similarly, the other boundary condition (at infinity) requires that $\varepsilon < 0$. Let us now substitute the assumed forms of the solutions into the second differential equation:

$$\varepsilon^2\delta e^{\varepsilon x} \sim -(\alpha + \beta e^{\gamma x})\varepsilon\delta e^{\varepsilon x}$$

After dividing by common factors, this gives

$$\varepsilon \sim -(\alpha + \beta e^{\gamma x}) \sim -\alpha$$

Because $\varepsilon < 0$, we must have $\alpha > 0$. Now let's substitute the assumed forms into the first differential equation:

$$3(\alpha + \beta e^{\gamma x})\gamma^2\beta e^{\gamma x} \sim 2(\gamma\beta e^{\gamma x} - \delta e^{-\alpha x})$$

Dividing this equation by $e^{\gamma x}$ yields

$$3(\alpha + \beta e^{\gamma x})\gamma^2\beta \sim 2(\gamma\beta - \delta e^{-(\alpha+\gamma)x}) \tag{3.24}$$

and, considering the behavior as $x \to \infty$, we find that we must have $(\alpha + \gamma) \geq 0$. If $(\alpha + \gamma) > 0$, then passing to the limit in (3.24) shows that

$$3\alpha\gamma^2\beta = 2\gamma\beta$$

and hence that $3\alpha\gamma = 2$. However, this is not possible because $\alpha > 0$ and $\gamma < 0$. We conclude then that $(\alpha + \gamma) = 0$, which is to say that $\gamma = -\alpha$. With this assumption we again pass to the limit in (3.24) to find that

$$3\alpha\gamma^2\beta = 2(\gamma\beta - \delta)$$

This is satisfied by the choice

$$\delta = -\alpha\beta - \tfrac{3}{2}\alpha^3\beta$$

Therefore, we have finally found an asymptotic solution of the ODEs with two free parameters $\alpha$ and $\beta$ that satisfies the boundary conditions at infinity:

$$y(x) \sim \alpha + \beta e^{-\alpha x}, \quad z(x) \sim -\left(\alpha\beta + \tfrac{3}{2}\alpha^3\beta\right)e^{-\alpha x}$$

In the program `ch3ex4.m`, we use terms in the power series through $O(x^{8/3})$ to approximate the solution component $y(x)$ for small $x$ and through $O(x^{11/3})$ to approximate $z(x)$. The leading terms in the expansions of the differential equations (effectively the residuals) are then both $O(x^2)$. The boundary conditions at the origin are replaced by the requirement that the numerical solutions agree with these series approximations at $x = \delta$. In this the coefficients $b$ and $c$ in the series are unknown parameters that are passed as components of a vector `P`. When written as a first-order system, there are four unknowns. The other unknowns, the derivatives $y'(x)$ and $z'(x)$, are approximated by the derivatives of the series for $y(x)$ and $z(x)$. The series approximation to the vector of unknowns is coded in a subfunction `series`. Besides its use in evaluating the residual in the boundary conditions, this function could be used to evaluate the solution anywhere in $[0, \delta]$. We could use the asymptotic solution with its unknown parameters at some large $\Delta$ just as we use the series at a small $\delta$ for the singularity at the origin. However, to illustrate a technique commonly used with codes that do not provide for unknown parameters, we first note that our asymptotic solutions satisfy

$$y(x) \sim \alpha, \quad z'(x) \sim -\alpha z(x)$$

From this we obtain a boundary condition

$$z'(x) \sim -y(x)z(x)$$

that does not involve an unknown parameter. The leading-term behavior of $y(x)$ at infinity is captured by the boundary condition $y'(\infty) = 0$ that also does not involve an unknown parameter. Although we must keep in mind the possibility that we might need to supply more information about the behavior of the solution at infinity if we are to solve this BVP, in the program `ch3ex4.m` we impose the boundary conditions

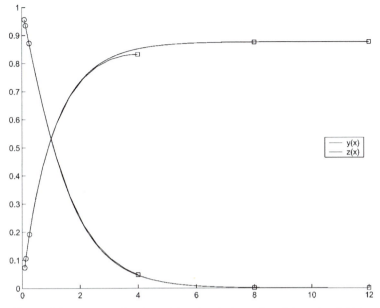

Figure 3.6: Solution of a BVP with a singularity at the origin and an infinite range.

$$y'(\Delta) = 0, \quad z'(\Delta) = -y(\Delta)z(\Delta)$$

which prove satisfactory for moderate values of $\Delta$. Exercise 3.9 considers a similar problem with singularities at both the origin and infinity.

It is straightforward to solve the BVP posed on $[\delta, \Delta]$. However, it is always prudent to vary the position of the end points to gain some confidence that they have been chosen appropriately. It is natural to choose the end points close to the singular points and large for infinity, but this can get us into just the kind of the trouble that our analysis is attempting to avoid – the solver may not be able to distinguish the various kinds of solutions of the ODEs. The program ch3ex4.m solves the BVP on three intervals and displays the solutions so that we can examine them for consistency. It makes use of a capability of the function bvpinit that is very convenient in these circumstances. After solving a problem on one interval, we can expect the solution to be a good guess for an interval that is only a little longer. As seen here, you can supply the solution structure and the longer interval to bvpinit in order to form a guess structure for the new interval. In this approach, the solution is extended automatically to the longer interval by extrapolation. How far you can extend a solution in this way depends on how rapidly it is changing, but it is also limited by the well-known hazards of polynomial extrapolation. The consistency of the solutions displayed in Figure 3.6 suggests that the final boundary points are close enough to the singular points that our approximate boundary conditions are acceptable.

```
function ch3ex4
% y(1) = y(x), y(2) = z(x), y(3) = y'(x), y(4) = z'(x)
global d
hold on
for i = 1:3
    D = 4*i; d = 1/D;
    if i == 1
        P = [-1; 0.5];  % Guess parameters P = [b; c]:
        solinit = bvpinit(linspace(d,D,5),[d; 0; 0; 0],P);
    else
        solinit = bvpinit(sol,[d,D]);
    end
    sol = bvp4c(@odes,@bcs,solinit);
    plot(sol.x,sol.y(1:2,:),sol.x(end),sol.y(1:2,end),'ro',...
        sol.x(1),sol.y(1:2,1),'ko');
    legend('y(x)','z(x)',0);
    axis([0 12 0 1]);
    drawnow
end
hold off

%===============================================================
function dydx = odes(x,y,P)
dydx = [ y(3); y(4); 2*(y(3) - y(2))/(3*y(1)); -y(1)*y(4) ];

function res = bcs(ya,yb,P)
global d
res = zeros(6,1);
res(1:4) = ya - series(d,P);
res(5:6) = [ yb(3); (yb(4) - (- yb(1)*yb(2))) ];

function y = series(x,P)
b = P(1); c = P(2);
yx  = x + b*x^(5/3) - c*x^2 - (5/7)*b^2*x^(7/3) ...
    + (7/6)*b*c*x^(8/3);
ypx = 1 + (5/3)*b*x^(2/3) - 2*c*x - (5/3)*b^2*x^(4/3)...
    + (28/9)*b*c*x^(5/3);
zx  = 1 + c*x - (1/6)*c*x^3 - (9/88)*b*c*x^(11/3);
zpx = c - (1/2)*c*x^2 - (3/8)*b*c*x^(8/3);
y = [ yx; zx; ypx; zpx];
```

Often the hardest part of solving a BVP is finding an initial estimate of the solution that is good enough to yield convergence. Here we have described solving the BVP on several intervals as a way of gaining confidence in the solution. It also serves as an example of *continuation*. The method of continuation exploits the fact that generally the solution of one BVP is a good guess for the solution of another with slightly different parameters. The program ch3ex4.m is an example of continuation in the length of the interval. If you have difficulty in finding a guess for the solution that is good enough to achieve convergence for the interval of interest, it is frequently the case that the problem is easier to solve on a shorter interval. The idea is then to solve a sequence of BVPs with the solution on one interval being used as a guess for the problem posed on a longer interval. If all goes well, you will reach the interval of interest with a guess that is good enough to yield convergence. Of course, you cannot use this technique to extend the interval *ad infinitum*; no matter how good your guess, eventually the solver will not be able to distinguish the different kinds of solutions. Further, continuation depends on the solution for one interval being much like that of an interval with nearly the same length, but for some BVPs this is not true. Particularly, continuation using the length of the interval as the continuation parameter can fail because sometimes a small difference in the interval length can change the situation from solving a BVP with a well-defined solution to attempting to solve a BVP that does not have a solution or has more than one.

## EXAMPLE 3.5.5

The Fisher BVP was used in Section 3.3.2 to illustrate the analysis of boundary conditions at infinity. The ODE is

$$U'' + cU' + U(1 - U) = 0$$

and the solution is to satisfy the boundary conditions

$$U(-\infty) = 1, \quad U(\infty) = 0$$

We found that the solution is not unique with these boundary conditions. After some analysis we specified boundary conditions that identify a unique solution by choosing a "large" number $Z$ and requiring that

$$\frac{U'(-Z)}{U(-Z) - 1} - \alpha = 0 \quad \text{for } \alpha = \frac{-c + \sqrt{c^2 + 4}}{2}$$

$$\frac{U(Z)}{e^{\beta Z}} - 1 = 0 \quad \text{for } \beta = \frac{-c + \sqrt{c^2 - 4}}{2}$$

In the program `ch3ex5.m` we use the variables `infty` for $Z$, and `alpha` and `beta` for $\alpha$ and $\beta$, respectively. The boundary conditions are coded in a straightforward way for the first-order system with two unknowns $y_1(z) = U(z)$ and $y_2(z) = U'(z)$. It is not difficult to solve this BVP once the boundary conditions are properly formulated. We use this problem to suggest some experiments that show what happens when the solver fails. We also solve the BVP for a range of wave speeds $c$. A study of this kind motivates our discussion of how to speed up the computations.

Measuring run times is a little tricky in MATLAB and quite naturally depends on the hardware and its configuration. All we want to do here is get a general idea of the relative effects of various options. The times we state were all obtained on a single-user computer as the elapsed time reported on the *second* invocation of

```
>> tic, ch3ex5; toc
```

With the default options specified by `options = []`, the program `ch3ex5.m` given at the end of this example ran in 5.76 seconds.

Like the MATLAB programs for solving stiff IVPs, by default `bvp4c` forms Jacobians using finite differences. Often it is easy to vectorize the evaluation of the ODE function $f(x, y)$ for a given value of $x$ and an array of vectors $y$. This can reduce the run time substantially, so the stiff IVP solvers have an option for this. It is a natural option for BVPs, too, but important additional gains are possible if the function is also vectorized with respect to the values of $x$. That is because, in discretizing the ODEs in the BVP context, all the values of the arguments are known in advance. Generalizing what is done for IVPs, there is an option for you to code the function $f(x, y)$ so that, when given a vector $x = [x_1, x_2, \dots]$ and a corresponding array of column vectors $y = [y_1, y_2, \dots]$, it returns an array of column vectors $[f(x_1, y_1), f(x_2, y_2), \dots]$. Vectorizing the ODE function in this way will often greatly reduce the run time. If the comment symbol is removed from the line in the program `ch3ex5.m` so that `Vectorized` is set to `on`, the run time is reduced to 3.40 seconds. This is a considerable reduction in a relative sense, showing that vectorization can be very advantageous. Yet the absolute run time is so short that it is scarcely worth the trouble. Still, vectorization is very little trouble for these ODEs. We might code the ODEs for scalar evaluation as

```
dydz = [ y(2); -(c*y(2) + y(1)*(1 - y(1))) ];
```

This is vectorized by changing the vectors to arrays and changing the multiplication to an array multiplication as shown in `ch3ex5.m`.

Most BVP solvers require you to provide analytical partial derivatives. Generally this is both inconvenient and unnecessary for the typical problem solved in MATLAB, so `bvp4c` does not. There may be good reasons for going to the trouble of providing analytical partial derivatives. One is that it may not be much trouble after all. For instance, linear ODEs can be written in the form

$$y'(x) = J(x)y(x) + q(x)$$

and it is generally convenient to supply a function for evaluating the Jacobian $J(x)$. Another reason is that BVPs are generally solved much faster with analytical partial derivatives. Although the `numjac` function that `bvp4c` uses to approximate partial derivatives numerically is very effective, the task is a difficult one and so analytical partial derivatives improve the robustness of the solver. For some difficult BVPs, supplying analytical partial derivatives is the difference between success and failure.

The code `bvp4c` permits you to supply analytical partial derivatives for either the ODEs or the boundary conditions or both. It is far more important to provide partial derivatives for the ODEs than the boundary conditions. You inform the solver that you have written a function for evaluating $\frac{\partial f}{\partial y}$ by providing its handle as the value of the `FJacobian` option. Similarly, you inform the solver of a function for evaluating analytical partial derivatives of the boundary conditions with the option `BCJacobian`. The function for evaluating the residual $g(ya, yb)$ in the boundary conditions involves two vectors, the approximate solution values $ya$ and $yb$ at the two ends of the interval. Accordingly, the function you write for evaluating partial derivatives of the boundary conditions must evaluate and return two matrices, $\frac{\partial g}{\partial ya}$ and $\frac{\partial g}{\partial yb}$.

A few comments about working out the Jacobian

$$J = \frac{\partial f}{\partial y} = \left(\frac{\partial f_i}{\partial y_j}\right)$$

and coding it as a function may be helpful. Suppose there are $d$ equations. If $J$ is sparse, you should initialize it to a sparse matrix of zeros with $J = $ `sparse(d,d)`. For each value of $i = 1, 2, \ldots, d$, you might then examine $f_i$ for the presence of components $y_j$. For each component that is present, you must work out the partial derivative of $f_i$ with respect to $y_j$ and then code the function to evaluate it as $J(i,j)$. Of course, you can evaluate the components of $J$ in any order that is convenient. This is a reasonable way to proceed even when you are treating $J$ as a dense matrix. The only difference is that you would initialize the matrix with $J = $ `zeros(d,d)` (or just $J = $ `zeros(d)`). The function you write for $f(x, y)$ must return a column vector $f$. A systematic way of working out partial derivatives when the matrix is not sparse is to proceed one column at a time. That is, for $j = 1, 2, \ldots, d$, work out the partial derivatives

$$J(:, j) = \frac{\partial f}{\partial y_j} = \left(\frac{\partial f_i}{\partial y_j}\right)$$

Once you have worked out expressions for all the partial derivatives, you can evaluate them in any convenient order. The MATLAB Symbolic Toolbox has a function `jacobian` that can be very helpful when working out partial derivatives for complicated functions. We illustrate its use with a script for the partial derivatives of the boundary conditions function of this example. We find it convenient to use names like `dBCdya` when providing analytical partial derivatives to remind ourselves which partial derivatives are being computed. With the help entry for `jacobian` and the variable names we have chosen, the program should be easy to follow.

```
syms res ya1 ya2 yb1 yb2 alpha beta infty
res = [ ya2/(ya1 - 1) - alpha
        yb1/exp(beta*infty) - 1 ];
dBCdya = jacobian(res,[ya1; ya2])
dBCdyb = jacobian(res,[yb1; yb2])
```

This program generates the (slightly edited) output

```
dBCdya = [ -ya2/(ya1-1)^2,        1/(ya1-1)]
         [              0,               0]

dBCdyb = [                 0,             0]
         [ 1/exp(beta*infty),             0]
```

A little work with a text editor turns this output into the subfunction `bcJac` of the program `ch3ex5.m`.

When the program `ch3ex5.m` was run with default values for all options except for `FJacobian`, the BVP was solved in 4.72 seconds, a fair improvement over the 5.76 seconds without this option. When `BCJacobian` was also supplied, the BVP was solved in 4.56 seconds. Sometimes vectorization reduces the run time more than supplying analytical Jacobians; other times, less. The options are independent, so you can do one or the other depending on what is convenient. Moreover, you can do both. In this instance, vectorizing the ODE function and supplying analytical partial derivatives for the ODE function and for the boundary conditions reduced the run time to 2.31 seconds. Clearly it is possible to reduce run times significantly if it is worth the trouble of supplying the solver with more information. And, it is worth the trouble more often than might seem likely at first. As with this example, if you expect to solve a problem repeatedly with different choices of parameters, then it might be worth making the effort even if an individual solution costs only a few seconds.

Now that we are solving the BVP efficiently, let's experiment with the solution process. Notice that in `ch3ex5.m` we specify a value of `infty` that depends on the wave speed $c$. It would seem natural simply to use a value large enough to solve the BVP for all the values of $c$ to be considered. However, if (say) you set `infty = 250` then the computation will fail with the message

```
?? Error using ==> bvp4c
Unable to refine the mesh any further --
the Jacobian of the collocation equations is singular
```

For the smaller values of $c$, the solution approaches its limit values rapidly – so rapidly that, with default tolerances, the solver cannot recognize the proper behavior at $Z = 250$. For instance, when $c = 5$,

$$U(250) \approx e^{250\beta} \approx 2.1846 \cdot 10^{-23}$$

This is too small for default tolerances, but by choosing `infty = 10*c` instead we obtain

$$U(50) \approx 2.9368 \cdot 10^{-5}$$

a value more comparable in size to the tolerances. When the various kinds of solution behavior cannot be distinguished in the discretized problem, the linear system is singular. When you receive this kind of error message, you need to consider whether you have the boundary conditions properly specified, especially when there is a singularity in the BVP. One reason for our analysis of asymptotic boundary conditions is to deal analytically with the solution where it is difficult to approximate numerically. Here we must take `infty` large enough that the asymptotic boundary condition properly describes the solution and yet not so large that we defeat the purpose of the asymptotic approximation.

In light of Example 3.5.4, it would seem natural to use the solution for one wave speed as a guess for the next. However, if we do this with

```
if i == 1
    solinit = bvpinit(linspace(-infty,infty,20),@guess);
else
    solinit = bvpinit(sol,[-infty,infty]);
end
```

we find that the run time increases to 10.60 seconds. It is not that continuation in the wave speed is a bad idea; often the only "simple" way to compute a solution on an infinite interval is to solve the problem repeatedly with modest increases in the length of the interval. What is happening here is that we have a guess that has the right asymptotic behavior at one end of the interval and the right qualitative behavior at the other. When we use the solution with a different wave number and a different interval as an estimate, we extrapolate the solution to a longer interval. The specified increases in length are acceptable in this instance, but we would be better off to use our asymptotic

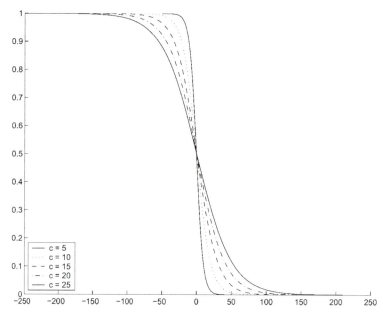

Figure 3.7: Solutions of the Fisher BVP for various wave speeds $c$.

expressions at the current wave speed because they provide a better solution estimate than does automatic polynomial extrapolation of the numerical solution for a different wave speed.

Estimates are provided for both the solution and the mesh. It is important to appreciate that, even when a function is supplied that approximates the solution everywhere in the interval, the solver uses its values only on the initial mesh. Figure 3.7 makes the point. The estimate coded in the program ch3ex5.m reflects the behavior of the desired solution at both end points of the interval, but it is a terrible approximation in the middle: it is 0 just to the left of the origin and 1 just to the right! It is interesting to experiment with the number of mesh points. When we specified an initial mesh of only five points, the solver failed because the asymptotic behavior is not revealed on a mesh of so few points. When we specified 200 equally spaced points, the solver was successful but the run time increased from 2.31 seconds to 5.77 seconds. Because the cost depends strongly on the number of mesh points, bvp4c tries to use as few as possible. Nevertheless, the goal is to solve the problem and not just to solve it with the fewest points possible, so for the sake of robustness bvp4c is cautious about discarding mesh points. Indeed, if the initial mesh and the guess for the solution on this mesh are satisfactory, it will not discard any. You might experiment with the effect on the run time of the choice of the number of equally spaced mesh points in the initial guess. In this experiment it would be illuminating to look at the number of mesh points (length(sol.x)) for the various values of the wave speed $c$. You will find that the solver does not need anything like 200 mesh points for most values of $c$. Another issue for this particular guess for the solution is that, as we increase the number of equally spaced points in the initial mesh, we place more points near the origin and hence the terrible approximation to the solution in this region has more influence on the computation. In Exercise 3.26 you are asked to explore the use of an asymptotic approximation (due to Murray) that provides a good estimate throughout the interval. Exercises 3.25, 3.27, and 3.28 ask that you experiment with vectorization and analytical partial derivatives.

```
function ch3ex5
global c alpha beta infty

options = [];
%options = bvpset(options,'Vectorized','on');
%options = bvpset(options,'FJacobian',@odeJac);
%options = bvpset(options,'BCJacobian',@bcJac);

color = [ 'r', 'b', 'g', 'm', 'k' ];
wave_speed = [5, 10, 15, 20, 25];
hold on
for i = 1:5
    c = wave_speed(i);
    alpha = (-c + sqrt(c^2 + 4))/2;
    beta  = (-c + sqrt(c^2 - 4))/2;
    infty = 10*c;
    solinit = bvpinit(linspace(-infty,infty,20),@guess);
    sol = bvp4c(@ode,@bc,solinit,options);
    plot(sol.x,sol.y(1,:),color(i));
    axis([-250 250 0 1]);
    drawnow
end
legend('c = 5', 'c = 10', 'c = 15', 'c = 20', 'c = 25',3);
hold off

%========================================================
function v = guess(z)
global c alpha beta infty
if z > 0
    v = [exp(beta*z); beta*exp(beta*z)];
else
```

```
      v = [(1 - exp(alpha*z)); -alpha*exp(alpha*z)];
end

function dydz = ode(z,y)
global c alpha beta infty
dydz =  [ y(2,:); -(c*y(2,:) + y(1,:).*(1 - y(1,:)))  ];

function dFdy = odeJac(z,y)
global c alpha beta infty
dFdy = [       0,            1
          (-1 + 2*y(1)),   - c ];

function res = bc(ya,yb)
global c alpha beta infty
res = [ ya(2)/(ya(1) - 1) - alpha
        yb(1)/exp(beta*infty) - 1 ];

function [dBCdya, dBCdyb] = bcJac(ya,yb)
global c alpha beta infty
dBCdya = [ -ya(2)/(ya(1) - 1)^2, 1/(ya(1) - 1)
                0                     0         ];

dBCdyb = [         0           0
            1/exp(beta*infty), 0 ];
```

## EXAMPLE 3.5.6

Example 1.4 of Ascher et al. (1995) describes flow in a long vertical channel with fluid injection through one side. The ODEs are

$$f''' - R[(f')^2 - ff''] + RA = 0$$
$$h'' + Rfh' + 1 = 0$$
$$\theta'' + P_e f\theta' = 0$$

Here $R$ is the Reynolds number and the Peclet number $P_e = 0.7R$. Because of the unknown parameter $A$, this system of total order 7 is subject to eight boundary conditions:

$$f(0) = f'(0) = 0, \quad f(1) = 1, \quad f'(1) = 1$$
$$h(0) = h(1) = 0, \quad \theta(0) = 0, \quad \theta(1) = 1$$

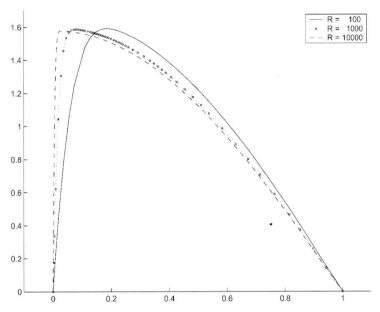

Figure 3.8: Solution of the fluid injection problem obtained by continuation.

It is quite common to study how a solution behaves as parameters change. For small changes in the parameters, we expect that the mesh and the numerical solution (and the unknown parameters, if present) for one set of parameters will be good guesses for the next set. The code bvp4c takes advantage of this by accepting a solution structure as a guess structure. This is not merely an efficient way to obtain solutions of BVPs for a range of parameter values, it is also an example of the approach to solving difficult BVPs by continuation. By a "hard" problem we mean here a BVP for which it is hard to find a guess that is good enough to result in convergence to an approximate solution. This particular BVP is also hard in the sense that, when $R$ is large, there is a boundary layer that requires many mesh points to resolve. Thus, it is easy to achieve convergence from a constant guess for a Reynolds number $R = 100$ but not for $R = 10,000$. Continuation shows us how the solution behaves as $R$ is changed and allows us to solve the BVP for relatively large values of $R$. When solving hard BVPs, often we require a large number of runs that often involve many iterations, so it may be worth going to some trouble to reduce the run time. We complement the discussion of this issue in Example 3.5.5 by showing what is different when there are unknown parameters.

Once you understand that the solution structure computed for one set of parameters can be used as the guess structure for another, it is easy to solve this BVP for several values of the Reynolds number $R$. For $R = 100$ we guess that all the components of the solution are constant with value 1 and that the unknown parameter $A = 1$ in the guess structure defined by the function `bvpinit`. Thereafter, the solution for one value of $R$ is used as guess for the next. The program `ch3ex6.m` plots the solutions as they are computed, resulting in Figure 3.8 (page 191), and the values for $A$ are displayed to the screen as

```
>> ch3ex6
For R =   100, A = 2.76.
For R =  1000, A = 2.55.
For R = 10000, A = 2.49.
```

This version of the program includes both vectorization and analytical partial derivatives.

```
function ch3ex6
global R

color = [ 'k', 'r', 'b' ];
options = bvpset('FJacobian',@Jac,'BCJacobian',@BCJac,...
                 'Vectorized','on');
R = 100;
sol = bvpinit(linspace(0,1,10),ones(7,1),1);
hold on
for i = 1:3
    sol = bvp4c(@ode,@bc,sol,options);
    fprintf('For R = %5i, A = %4.2f.\n',R,sol.parameters);
    plot(sol.x,sol.y(2,:),color(i));
    axis([-0.1 1.1 0 1.7]);
    drawnow
    R = 10*R;
end
legend('R =    100','R =   1000','R = 10000',1);
hold off


%================================================================
function dydx = ode(x,y,A);
global R
P = 0.7*R;
dydx = [ y(2,:); y(3,:); R*(y(2,:).^2- y(1,:).*y(3,:)-A);...
         y(5,:); -R*y(1,:).*y(5,:)-1; y(7,:); -P*y(1,:).*y(7,:) ];

function [dFdy,dFdA] = Jac(x,y,A)
global R
```

```
dFdy = [        0,         1,         0,      0,   0,      0,         0
                0,         0,         1,      0,   0,      0,         0
          -R*y(3),  2*R*y(2),  -R*y(1),      0,   0,      0,         0
                0,         0,         0,      0,   1,      0,         0
          -R*y(5),         0,         0,      0, -R*y(1), 0,         0
                0,         0,         0,      0,   0,      0,         1
    -7/10*R*y(7),         0,         0,      0,   0,      0, -7/10*R*y(1) ];

dFdA = [ 0; 0; -R; 0; 0; 0; 0];

function res = bc(ya,yb,A)
res = [ ya(1); ya(2); yb(1)-1; yb(2);...
        ya(4); yb(4); ya(6); yb(6)-1 ];

function [dBCdya,dBCdyb,dBCdA] = BCJac(ya,yb,A)
dBCdya = [ 1, 0, 0, 0, 0, 0, 0
           0, 1, 0, 0, 0, 0, 0
           0, 0, 0, 0, 0, 0, 0
           0, 0, 0, 0, 0, 0, 0
           0, 0, 0, 1, 0, 0, 0
           0, 0, 0, 0, 0, 0, 0
           0, 0, 0, 0, 0, 1, 0
           0, 0, 0, 0, 0, 0, 0 ];

dBCdyb = [ 0, 0, 0, 0, 0, 0, 0
           0, 0, 0, 0, 0, 0, 0
           1, 0, 0, 0, 0, 0, 0
           0, 1, 0, 0, 0, 0, 0
           0, 0, 0, 0, 0, 0, 0
           0, 0, 0, 1, 0, 0, 0
           0, 0, 0, 0, 0, 0, 0
           0, 0, 0, 0, 0, 1, 0 ];

dBCdA = zeros(8,1);
```

Providing analytical partial derivatives is more complicated for this BVP than that of Example 3.5.5 because, when there are unknown parameters, we must also provide partial derivatives with respect to these parameters. We chose names for the variables in the program `ch3ex6.m` that indicate what must be returned by the subfunction `Jac`. The array `dFdy` is the usual Jacobian matrix of the ODEs, here a $7 \times 7$ matrix. When there are unknown parameters, there is a second output argument that is the matrix of partial derivatives of the ODEs with respect to the unknown parameters. Here there are seven ODEs and one unknown parameter $A$, so the array dFdA is a $7 \times 1$ vector. In general, if there are $d$ equations and $m$ unknown parameters $p_j$, then the matrix of partial derivatives with respect to the parameters is $d \times m$. A systematic way to work them out by hand is to form column $j$ of the matrix as the partial derivative of $f$ with respect to the parameter $p_j$. As we saw in Example 3.5.5, the MATLAB Symbolic Toolbox has a function jacobian that can help generate the matrices of partial derivatives. We might, for example, use

```
syms y y1 y2 y3 y4 y5 y6 y7 R A P F
y = [ y1; y2; y3; y4; y5; y6; y7];
P = 0.7*R;
F = [ y2; y3; R*(y2^2- y1*y3-A); y5;
      -R*y1*y5-1; y7; -P*y1*y7 ];
dFdy = jacobian(F,y)
dFdA = jacobian(F,A)
```

and edit the output a little to obtain the subfunction Jac. The boundary conditions are handled similarly. The array dBCdya contains the partial derivatives of the residual vector $g(ya, yb)$ with respect to the argument $ya$. Here there are eight residuals and $ya$ has seven components, so the array dBCdya is an $8 \times 7$ matrix and the same is true of the array dBCdyb corresponding to $yb$. In general, if there are $d$ equations and $m$ parameters, these matrices are $(d + m) \times d$. There is a third output argument, which is the matrix of partial derivatives of the residual with respect to the unknown parameters. Here there is only one unknown parameter, so the array dBCdA is $8 \times 1$. In general this matrix is $(d + m) \times m$. In solving BVPs with analytical partial derivatives, you might sometimes receive an error message stating that the dimensions of some arrays internal to bvp4c are not properly matched. Generally this means that the size of one of your partial derivative matrices is wrong. Compare the sizes of the arrays you are computing to the general expressions just stated. For this example, most of the entries of the various matrices of partial derivatives are zero and so, if we had coded them by hand, we would have initialized the matrices to zero and then evaluated only the nonzero entries. However, we generated the matrices for this example using the Symbolic Toolbox and just edited the output with its many zeros to obtain the subfunctions of ch3ex6.m.

By measuring run times as in Example 3.5.5, we found that with default options the program ch3ex6.m ran in 13.35 seconds. It is quite easy to vectorize the evaluation of the ODEs. Just as for Example 3.5.5, this is a matter of changing scalar quantities like y(1) into arrays like y(1,:) and changing from scalar operations to array operations by replacing * and ^ with .* and .^, respectively. Vectorizing the ODE function reduced the run time to 8.02 seconds. When using analytical partial derivatives with Vectorized set to off, the BVP was solved in 6.64 seconds. With analytical partial derivatives and Vectorized set to on, the run time was reduced to 4.01 seconds.

Continuing on to Reynolds number $R = 1,000,000$, the solver bvp4c reports

```
Warning: Unable to meet the tolerance without using more
than 142 mesh points.
```

Generally you do not need to concern yourself with storage issues, but the solver does have a limit on the number of mesh points. This is set by default to floor(1000/d), where d is the number of ODEs. The value is somewhat arbitrary and, if you should receive this warning, you can use the option Nmax to increase it. You can receive this warning because the guesses for the mesh and solution are not good enough and the solver has tried unsuccessfully to achieve convergence by increasing the number of mesh points. Or, the solution may be hard to approximate and the solver just needs more mesh points to represent it to the specified accuracy. Or, there may be no solution at all! Here, when we increase Nmax to 500, the code is able to solve the BVP. This is a hard problem for bvp4c. It needed 437 mesh points to approximate the solution even with default tolerances. And even with vectorization and with analytical partial derivatives, the computation took 22.69 seconds.

## EXAMPLE 3.5.7

Continuation is an extremely important tool for the practical solution of BVPs. We have seen examples of continuation in the length of the interval. This is certainly natural when solving problems on infinite intervals, but it can be valuable when the interval is finite, too. We have seen an example of continuation in physical parameters. Often this is natural because you are interested in solutions for more than one set of parameters, but sometimes it is useful to vary a quantity because you can solve the BVP easily for one value of the quantity. This is closely related to a general form of continuation in an artificial parameter that we take up in this example. Suppose that you are having trouble solving the BVP

$$y' = f(x, y), \quad 0 = g(y(a), y(b))$$

because you cannot find a guess that is good enough for the solver to converge. It is not unusual that a simplified model BVP

$$y' = F(x, y), \quad 0 = G(y(a), y(b))$$

can be solved without difficulty, perhaps even analytically. Often it is useful to approximate the given problem by a linear one because for such problems there is no iteration in solvers like bvp4c. The idea is to continue from the solution of the easy problem to the solution of the problem you want to solve. One way to do this is to introduce an artificial parameter $\mu$ and solve the family of BVPs

$$y' = \mu f(x, y) + (1 - \mu)F(x, y)$$
$$0 = \mu g(y(a), y(b)) + (1 - \mu)G(y(a), y(b))$$

for $\mu$ ranging from 0 to 1. This plausible approach to solving difficult BVPs is very useful in practice, but it is not always successful. Finding an easy problem that captures the behavior of the original problem may be crucial to success. Related to this is the possibility that continuation might fail because there is a value of $\mu$ for which the corresponding BVP simply does not have a solution. Exercise 3.30 provides an example. How much you change $\mu$ at each step in the continuation process can mean the difference between success and failure.

Chapter 7 of Roberts & Shipman (1972) is devoted to continuation. To illustrate the technique, we follow their discussion of a BVP in Examples 1 and 5 of that chapter. The ODEs are

$$y_1' = y_2$$
$$y_2' = y_3$$
$$y_3' = -\left(\frac{3 - n}{2}\right)y_1 y_3 - ny_2^2 + 1 - y_4^2 + sy_2$$
$$y_4' = y_5$$
$$y_5' = -\left(\frac{3 - n}{2}\right)y_1 y_3 - (n - 1)y_2 y_4 + s(y_4 - 1)$$

Here the parameters $n = -0.1$ and $s = 0.2$. The ODEs are to be solved on $[0, b]$ with boundary conditions

$$y_1(0) = 0, \quad y_2(0) = 0, \quad y_4(0) = 0, \quad y_2(b) = 0, \quad y_4(b) = 1$$

for $b = 11.3$. Let's write the ODEs as the sum of their linear terms and a multiple $\delta = 1$ of their nonlinear terms:

$$y' = \begin{pmatrix} y_2 \\ y_3 \\ 1 + sy_2 \\ y_5 \\ s(y_4 - 1) \end{pmatrix} + \delta \begin{pmatrix} 0 \\ 0 \\ -\left(\frac{3-n}{2}\right)y_1 y_3 - ny_2^2 - y_4^2 \\ 0 \\ -\left(\frac{3-n}{2}\right)y_1 y_3 - (n - 1)y_2 y_4 \end{pmatrix}$$

With the given boundary conditions and $\delta = 0$ in these ODEs, we have an approximating linear BVP that can be solved with a nominal guess. We then use the solution for one value of $\delta$ as the guess when solving the BVP with a larger value of $\delta$. We continue in this way until we arrive at $\delta = 1$ and the solution of the BVP that interests us. The program ch3ex7.m provides the details.

```
function sol = ch3ex7
global delta

sol = bvpinit(linspace(0,11.3,5),ones(5,1));
for delta = [0 0.1 0.5 1]
    sol = bvp4c(@odes,@bcs,sol);
%    plot(sol.x,sol.y)
%    drawnow
%    pause
end
plot(sol.x,sol.y)
axis([0 11.3 -2 1.5])
fprintf('Reference values: y_3(0) = -0.96631, y_5(0) =  0.65291\n')
fprintf('Computed values:  y_3(0) = %8.5f, y_5(0) = %8.5f\n',...
        sol.y(3,1),sol.y(5,1))

%====================================================
function dydt = odes(t,y)
global delta
n = -0.1;
s = 0.2;
c = -(3- n)/2;
linear = [ y(2); y(3); 1+s*y(2); y(5); s*(y(4) - 1) ];
nonlinear = [ 0; 0; (c*y(1)*y(3) - n*y(2)^2 - y(4)^2); ...
            0; (c*y(1)*y(5) - (n-1)*y(2)*y(4))          ];
dydt = linear + delta*nonlinear;

function res = bcs(ya,yb)
res = [ ya(1); ya(2); ya(4); yb(2); yb(4)-1 ];
```

The program displays a comparison of two values of the computed solution and reference values reported by Roberts & Shipman (1972):

```
>> sol = ch3ex7;
Reference values: y_3(0) = -0.96631, y_5(0) =  0.65291
Computed values:  y_3(0) = -0.96629, y_5(0) =  0.65293
```

The agreement is quite satisfactory for the default tolerances used in the computation. The program also displays all the solution components, as shown here in Figure 3.9. The lines that have been commented out in ch3ex7.m let you see how the solution of the BVP changes with $\delta$. As it turns out, the linear approximating BVP of $\delta = 0$ does not provide

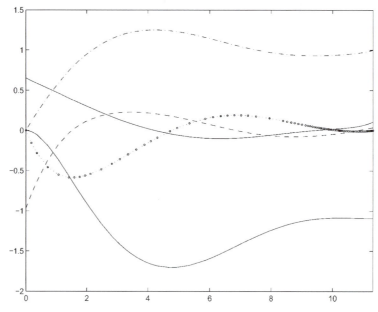

Figure 3.9: BVP solved by continuation.

a very good approximation to the solution when $\delta = 1$, but it is good enough that convergence is obtained with relatively large changes in $\delta$ at each step of the continuation process.

## EXAMPLE 3.5.8

Some solvers provide for separated boundary conditions specified at more than two points – that is, separated multipoint boundary conditions. Among them are COLNEW (Bader & Ascher 1987) and others that adopted the COLSYS (Ascher et al. 1995) user interface. Many solvers (including bvp4c) do not, so we illustrate here one way to prepare multipoint BVPs for solution with such codes. Other ways of doing this and examples of multipoint BVPs are found in Ascher & Russell (1981) and Ascher et al. (1995, chap. 11).

Chapter 8 of Lin & Segel (1988) is devoted to the study of a physiological flow problem. After considerable preparation, Lin and Segel arrive at equations that can be written for $0 \le x \le \lambda$ as

$$v' = \frac{C-1}{n}$$

$$C' = \frac{vC - \min(x, 1)}{\eta}$$

Here $n$ and $\eta$ are dimensionless (known) parameters and $\lambda > 1$. The boundary conditions are

$$v(0) = 0, \quad C(\lambda) = 1$$

The quantity of most interest is the dimensionless emergent osmolarity defined by

$$O_s = \frac{1}{v(\lambda)}$$

Using perturbation methods, Lin and Segel approximate it for small $n$ by

$$O_s \approx \frac{1}{1 - K}$$

where

$$K = \frac{\lambda \sinh(\kappa/\lambda)}{\kappa \cosh(\kappa)}$$

and the parameter $\kappa$ is such that

$$\eta = \frac{\lambda^2}{n\kappa^2}$$

The term $\min(x, 1)$ in the equation for the derivative $C'(x)$ is not smooth at $x = 1$. Indeed, Lin and Segel describe this BVP as two problems, one set on the interval $[0, 1]$ and the other on the interval $[1, \lambda]$, connected by the requirement that the solution components $v(x)$ and $C(x)$ be continuous at $x = 1$. Numerical methods have less than their usual order of convergence when the ODEs, and hence their solutions, are not smooth. Despite this, `bvp4c` is sufficiently robust that it can solve the problem formulated in the way that Lin and Segel suggest without difficulty. This is certainly the easier way to solve this particular problem, but it is better practice to recognize that this is a multipoint BVP. In particular, this problem is a three-point BVP because it involves boundary conditions at three points rather than at the two that we have seen in all earlier examples. A standard way to reformulate this multipoint BVP as a two-point BVP is first to introduce unknowns $y_1(x) = v(x)$ and $y_2(x) = C(x)$ for the interval $0 \leq x \leq 1$, so that the differential equations there are

$$\begin{aligned}
\frac{dy_1}{dx} &= \frac{y_2 - 1}{n} \\
\frac{dy_2}{dx} &= \frac{y_1 y_2 - x}{\eta}
\end{aligned} \tag{3.25}$$

One of the boundary conditions becomes $y_1(0) = 0$. Next, unknowns $y_3(x) = v(x)$ and $y_4(x) = C(x)$ are introduced for the interval $1 \leq x \leq \lambda$, resulting in the equations

$$\frac{dy_3}{dx} = \frac{y_4 - 1}{n}$$

$$\frac{dy_4}{dx} = \frac{y_3 y_4 - 1}{\eta}$$

The other boundary condition becomes $y_4(\lambda) = 1$. With these new variables, the continuity conditions on the solution components $v$ and $C$ become $y_1(1) = y_3(1)$ and $y_2(1) = y_4(1)$. This is all easy enough, but the trick is to solve the four differential equations simultaneously. This is accomplished by defining a new independent variable

$$\tau = \frac{x - 1}{\lambda - 1}$$

for the interval $1 \leq x \leq \lambda$. Like the independent variable $x$ in the first interval, this independent variable ranges from 0 to 1 in the second interval. In terms of this new independent variable, the differential equations on the second interval become

$$\frac{dy_3}{d\tau} = \frac{(\lambda - 1)(y_4 - 1)}{n}$$

$$\frac{dy_4}{d\tau} = \frac{(\lambda - 1)(y_3 y_4 - 1)}{\eta} \tag{3.26}$$

The boundary condition $y_4(x = \lambda) = 1$ becomes $y_4(\tau = 1) = 1$. The continuity condition $y_1(x = 1) = y_3(x = 1)$ becomes $y_1(x = 1) = y_3(\tau = 0)$. Similarly, the other continuity condition becomes $y_2(x = 1) = y_4(\tau = 0)$. Since the differential equations for the four unknowns are connected only through the boundary conditions and since both sets are to be solved for an independent variable that is ranging from 0 to 1, we can combine the ODEs in (3.25) and (3.26) as one system to be solved for the interval $0 \leq t \leq 1$. In the common independent variable $t$, the boundary conditions are

$$y_1(0) = 0, \quad y_4(1) = 1, \quad y_1(1) = y_3(0), \quad y_2(1) = y_4(0)$$

That the boundary conditions arising from requiring continuity are nonseparated does not matter when using `bvp4c`, but solvers that do not accept either the problem in its original multipoint form or nonseparated boundary conditions will require additional preparation of the problem to separate the boundary conditions (cf. Example 3.5.3).

The program `ch3ex8.m` solves the three-point BVP for $\kappa = 2, 3, 4$, and 5 when the parameters $n = 5 \cdot 10^{-2}$ and $\lambda = 2$. The solution for one value of $\kappa$ is used as guess for the next – another example of continuation in a physical parameter. For each value of $\kappa$, the computed osmolarity $O_s$ is compared to the approximation given by Lin and Segel. The known parameters are passed through `bvp4c` as additional arguments. Because the BVP is solved with default options, `[ ]` is used as a placeholder. The only remaining complication is that, in order to plot a solution, we must undo the change of variables in the second

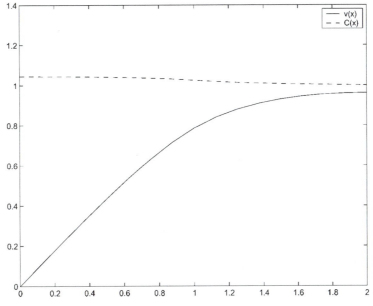

Figure 3.10: Solution of a three-point BVP.

interval and assemble a solution for the whole interval $[0, \lambda]$. The program `ch3ex8.m` displays

| kappa | computed Os | approximate Os |
|-------|-------------|----------------|
| 2 | 1.462 | 1.454 |
| 3 | 1.172 | 1.164 |
| 4 | 1.078 | 1.071 |
| 5 | 1.039 | 1.034 |

and plots the solution components $v(x)$ and $C(x)$ for $\kappa = 5$, as seen in Figure 3.10.

```
function ch3ex8

sol = bvpinit(linspace(0,1,5),[1 1 1 1]);
n = 5e-2;
lambda = 2;
fprintf(' kappa    computed Os   approximate Os\n')
for kappa = 2:5
  eta = lambda^2/(n*kappa^2);
  sol = bvp4c(@odes,@bcs,sol,[],n,lambda,eta);
  K2 = lambda*sinh(kappa/lambda)/(kappa*cosh(kappa));
```

```
  approx = 1/(1 - K2);
  computed = 1/sol.y(3,end);
  fprintf('  %2i    %10.3f    %10.3f\n',kappa,computed,approx);
end


% v and C are computed separately on 0 <= x <= 1 and 1 <= x <= lambda.
% A change of independent variable is used for the second interval.
% First it must be undone to obtain the corresponding mesh in x and
% then a solution assembled for all of 0 <= x <= lambda.
x = [sol.x sol.x*(lambda-1)+1];
y = [sol.y(1:2,:) sol.y(3:4,:)];
plot(x,y(1,:),x,y(2,:))
legend('v(x)','C(x)')


%=================================================
function dydx = odes(x,y,n,lambda,eta)
dydx = [ (y(2) - 1)/n
         (y(1)*y(2) - x)/eta
         (lambda - 1)*(y(4) - 1)/n
         (lambda - 1)*(y(3)*y(4) - 1)/eta ];


function res = bcs(ya,yb,n,lambda,eta)
res = [ ya(1); yb(4)-1; yb(1)-ya(3); yb(2)-ya(4)];
```

### ■ EXERCISE 3.15

Verify that the BVP consisting of the ODE

$$y'' + \lambda e^y = 0$$

with boundary conditions $y(0) = 0 = y(1)$ has the first integral (conservation law)

$$(y'(x))^2 + 2\lambda(e^{y(x)} - 1) = (y'(0))^2$$

*Hint:* Multiply the ODE by $2y'(x)$ and integrate. A first integral can be used to verify a numerical solution. As discussed in Section 1.5, this must be interpreted carefully: If the numerical solution does not satisfy a conservation law very well, you can be sure that it is not a very accurate solution to the BVP. On the other hand, even if the numerical solution satisfies the conservation law well, you cannot thereby conclude that it is an accurate solution – it may be, but it may have large errors that are correlated in such a way that the law is satisfied well. For the choice $\lambda = 1$, the program ch3ex1.m computes approximations Sxint(1,:) to the solution $y(x)$ and Sxint(2,:) to its derivative $y'(x)$ at the values $x$ of `xint`. Using vector operations, evaluate and plot the residual obtained by substituting the numerical solution into the conservation law. Is the size of the residual about what you might expect for the default tolerances used?

### ■ EXERCISE 3.16

A BVP is discussed in Chapter 1 that has two solutions displayed in Figure 1.3. Reproduce this figure by solving the equations for the motion of the cannon shot,

$$y' = \tan(\phi)$$

$$v' = -\frac{g \sin(\phi) + \nu v^2}{v \cos(\phi)}$$

$$\phi' = -\frac{g}{v^2}$$

For the figure, the parameters $g = 0.032$ and $\nu = 0.02$, the interval is $[0, 5]$, and the muzzle velocity $v(0) = 0.5$. The other two boundary conditions are that the shot starts and ends at ground level, that is, $y(0) = 0 = y(5)$. Use constant guesses for the two solutions. Use the function `deval` to evaluate the solutions at the points `linspace(0,5)` so that you plot smooth graphs.

### ■ EXERCISE 3.17

This problem, which models a tubular reactor with axial dispersion, was studied by Finlayson (1972, sec. 5.4). An isothermal situation with an irreversible reaction of order $n$ leads to the ODE

$$y'' = P_e(y' + Ry^n)$$

Here $P_e$ is the axial Peclet number and $R$ is the reaction rate group. The boundary conditions are

$$y'(0) = P_e(y(0) - 1), \quad y'(1) = 0$$

Using an orthogonal collocation method, Finlayson finds that $y(0) = 0.63678$ and $y(1) = 0.45759$ when $P_e = 1$, $R = 2$, and $n = 2$. These values are consistent with those obtained by other workers using a finite difference method. Solve this problem with an initial choice of ten equally spaced points in the interval and constant guesses of 0.5 for $y(x)$ and 0 for $y'(x)$. Plot the numerical solution and compare its values at $x = 0$ and $x = 1$ to those obtained by Finlayson.

### ■ EXERCISE 3.18

Example 5.3 of Bailey et al. (1968) considers a long, thin cantilever beam of length $L$ and flexural rigidity $B$ subjected to a concentrated vertical load $P$ at the free end. The displacement of the beam can be described in terms of arc length $s$ and the angle $\phi(s)$ that the beam makes with the horizontal. This angle is determined by the differential equation

$$\frac{d^2\phi}{ds} + \frac{P}{B}\cos(\phi) = 0$$

and the boundary conditions

$$\phi(0) = 0, \quad \phi'(L) = 0$$

In Bailey et al. (1968) the solution of this BVP is used to obtain a quantity of physical interest, but here you are to plot the deflected beam. For that you will need to add the equations

$$\frac{dx}{ds} = \cos(\phi), \quad \frac{dy}{ds} = -\sin(\phi)$$

and initial conditions

$$x(0) = 0, \quad y(0) = 1$$

Solve this BVP with the nominal values $L = 10$ and $P/B = 0.001$, and plot $(x(s), y(s))$. This is an easy problem that can be solved with constant guesses for the unknowns.

### ■ EXERCISE 3.19

Exercise 1.9 considers how to write in standard form a BVP used by Caughy (1970) to describe the large amplitude whirling of an elastic string. Here we consider how to solve the BVP numerically. Holmes (1995, pp. 32–3) writes the solution $\mu(x)$ as $\varepsilon y(x)$ and then approximates $y(x)$ for "small" $\varepsilon$ using perturbation theory. The function $y(x)$ satisfies the ODE

$$y'' + \omega^2\left(\frac{1-\alpha^2}{H}\frac{1}{\sqrt{1+\varepsilon^2 y^2}} + \alpha^2\right)y = 0$$

and boundary conditions

$$y'(0) = 0, \quad y'(1) = 0$$

Here $\alpha$ is a physical constant with $0 < \alpha < 1$. Because the whirling frequency $\omega$ is to be determined as part of solving the BVP, there is another boundary condition

$$y(0) = 1$$

An unusual aspect of this problem is that the constant $H$ is defined in terms of the solution $y(x)$ throughout the interval:

$$H = \frac{1}{\alpha^2}\left[1 - (1-\alpha^2)\int_0^1 \frac{dx}{\sqrt{1+\varepsilon^2 y^2(x)}}\right]$$

Following the suggestions of Exercise 1.9, formulate this BVP in standard form. Solve it numerically for $\alpha = 0.5$ and $\varepsilon = 1$. Show analytically that, when $\varepsilon = 0$, there is a solution $y(x) = \cos(\pi x)$ with $\omega = \pi$. Use this solution and derived quantities as guesses for the numerical solution when $\varepsilon > 0$. If this were a more difficult BVP, you might need to use continuation in the parameter $\varepsilon$ to solve the problem with $\varepsilon = 1$, but you will find that the guesses provided by the solution for $\varepsilon = 0$ are good enough to yield convergence for $\varepsilon = 1$. Plot $y(x)$ and report $\omega$. Caughy shows analytically that $y(0.5) = 0$ and $y(1) = -1$. What is the value of your numerical solution at these points?

### ■ EXERCISE 3.20

Example 6 of Kubiček, Hlaváček, & Holodnick (1979) describes the concentration and temperature fields in a tubular reactor with recirculation by the ODEs

$$y' = D_a(1 - y)\exp\left(\frac{\gamma\theta}{\gamma + \theta}\right)$$

$$\theta' = BD_a(1 - y)\exp\left(\frac{\gamma\theta}{\gamma + \theta}\right) - \beta(\theta - \theta_c)$$

with nonseparated boundary conditions

$$y(0) = (1 - \lambda)y(1), \quad \theta(0) = (1 - \lambda)\theta(1)$$

When the parameters have the values

$$\beta = 0, \quad \gamma = 20, \quad \lambda = 0.5, \quad \theta_c = 1, \quad B = 6, \quad D_a = 0.05$$

Kubiček et al. report that the initial values $(y(0), \theta(0)) \approx (0.1, 0.6)$. Confirm this and plot the solution. (You might try a guess of [1; 1] for $(y, \theta)$.) They also report that if the parameter $D_a$ is changed to 0.053 and the other parameters retain their values, then there are three solutions with initial values $(y(0), \theta(0))$ that are approximately equal to $(0.1, 0.7)$, $(0.3, 1.8)$, and $(0.44, 2.6)$. To gain experience with computing multiple solutions, try solving for all three. (You can solve all these problems using constant guesses for the solution components.)

### ■ EXERCISE 3.21

Edwards (1997) uses Mathematica to discuss a problem in Newton's *Principia Mathematica* "concerning the shape of a solid of revolution that experiences minimal resistance to rapid motion through a 'rare medium' consisting of elastic particles". A BVP for the shape is obtained using the calculus of variations and solved by a shooting method using NDSolve, the numerical IVP solver of the Mathematica PSE. The ODE is

$$y''(t) = \frac{y'(t)[1 + (y'(t))^2]}{t[3(y'(t))^2 - 1]}$$

and the boundary conditions are

$$y(t_0) = 0, \quad y'(t_0) = 1, \quad y(1) = 1$$

There are three boundary conditions because $t_0 \in [0, 1]$ is an unknown parameter. Edwards finds that $t_0 \approx 0.3509$. The nose cone is a cylinder of revolution about the vertical axis and so, with $t_0 > 0$, the tip of the cone is flat: a disk of radius $t_0$. Edwards also finds that the reduced drag coefficient

$$k = t_0^2 + \int_{t_0}^{1} \frac{2\tau}{(y'(\tau))^2 + 1} \, d\tau$$

is about 0.3748. You are to confirm these two numerical results by solving the BVP with bvp4c. Begin by using the variables $y_1(t) = y(t)$ and $y_2(t) = y'(t)$ to write the BVP as a first-order system. As in Section 1.3, introduce a variable $y_3(t)$ for the computation of $k$. A natural choice has $y_3(1) = 1$ and a simple ODE for $y_3(t)$ that involves only $t$ and $y_2(t)$. With this choice you have $k = y_3(t_0)$ after solving the BVP. As in Example 3.5.3, you now need to change the independent variable so as to have a fixed interval. Let $x$ be an independent variable that ranges from 0 to 1 as $t$ ranges from $t_0$ to 1. With this new independent variable you formulate a BVP defined on $[0, 1]$ that involves an unknown parameter $t_0$. Solve this BVP with default tolerances. Edwards guessed that $t_0 \approx 0.5$. It would be plausible to guess that $y(x) \approx x$. After solving the BVP, plot $y(t)$ on $[0, 1]$ to see the shape of the nose cone. You might enjoy plotting the cone as a surface.

■ **EXERCISE 3.22**
In Chapter 1, the BVP consisting of the equation $\theta'' + \sin(\theta) = 0$ (i.e., equation (1.6)) and boundary conditions $\theta(0) = 0$ and $\theta(+\infty) = \pi$ arose when discussing the motion of a pendulum. The solution of this BVP is the dotted curve of Figure 1.2. Solve this BVP yourself by replacing the boundary condition at infinity with $\theta(T) = \pi$. Solve the BVP for several choices of $T$ (e.g., $T = 5$, 10, and 15) to gain confidence in your solution. By analytical means it was determined in Chapter 1 that the initial slope $\theta'(0) = 2$. Confirm the accuracy of your numerical solution by comparing this value to the one you compute for the initial slope.

■ **EXERCISE 3.23**
Cebeci & Keller (1971) use shooting methods to solve the Falkner–Skan problem that arises from a similarity solution of viscous, incompressible, laminar flow over a flat plate. The ODE

$$f''' + ff'' + \beta(1 - (f')^2) = 0$$

is solved subject to the boundary conditions

$$f(0) = 0, \quad f'(0) = 0, \quad f'(+\infty) = 1$$

It appears that physically relevant solutions exist only for $-0.19884 \leq \beta \leq 2$. Cebeci and Keller deal with the boundary condition at infinity by imposing it at a finite point. For a range of $\beta > 0$, the BVP can be solved in a straightforward way with a shooting method, though continuation in $\beta$ is needed to compute solutions for some values of the parameter. The value $\beta = 0.5$ was relatively difficult for their shooting code, but you will find that the BVP with this value of $\beta$ and the boundary condition $f'(6) = 1$ is not difficult for bvp4c. Considering that $f'(x) \approx 1$ on the last part of the interval, it would be reasonable to guess that $f(x) \approx x$ and $f''(x) \approx 0$. Verify from a plot of your solution that $f'(x) \rightarrow 1$ quickly as $x$ increases, making it plausible that $x = 6$ is large enough to serve as infinity in the boundary condition. Compare the value that you compute for $f''(0)$ to the value 0.92768 reported by Cebeci and Keller.

### ■ EXERCISE 3.24

In a discussion of viscous incompressible flow past a semi-infinite body, Cole (1968, p. 159) considers the BVP

$$G' - xF' = 0, \quad G(0) = F(0) = 0$$

$$F'' + 2(xF - G)F' + 2(1 - F^2) = 0, \quad F(+\infty) = 1$$

Solve this problem by replacing the boundary condition at infinity with the boundary condition $F(X) = 0$ for some finite value $X$. Several values should be tried to gain confidence in your solution, for example, $X = 2$, 3, and 4. The quantity of most physical interest (the skin friction) is a multiple of $F'(0)$, so report the values that you compute for $F'(0)$.

This BVP can be studied much like that of Exercise 3.8. If we suppose that $G(x)$ has a limit as $x \rightarrow \infty$, or at least that it is bounded, then we can use the boundary condition $F(\infty) = 1$ to conclude that $xF - G \sim x$. Using this approximation and neglecting the term $(1 - F^2)$ that is small for large $x$, we are led to approximate the second ODE by $F'' + 2xF' = 0$. Solving this ODE, we find that

$$F'(x) \sim \beta e^{-x^2}$$

for a constant $\beta$. Integrating and imposing the boundary condition at infinity, we then find that

$$F(x) \sim 1 - \beta \int_x^\infty e^{-t^2}\, dt = 1 - \frac{\beta \sqrt{\pi}}{2} \operatorname{erfc}(x)$$

Using the standard asymptotic representation of the complementary error function,

$$\operatorname{erfc}(x) \sim \frac{e^{-x^2}}{x\sqrt{\pi}}$$

we find that $F(x) \to 1$ very quickly as $x$ increases. Turning now to the first ODE, we have

$$G'(x) = xF'(x) \sim \frac{\beta}{2}(2xe^{-x^2})$$

which we integrate to obtain

$$G(x) \sim G(\infty) - \frac{\beta}{2}e^{-x^2}$$

Not only does $G(x)$ have a limit as $x \to \infty$, it approaches the limit very quickly. Because both the unknown functions approach limits very quickly as $x$ increases, we can – indeed, must – impose the numerical boundary condition $F(X) = 1$ at what seems like a very small value of $X$. This BVP is easily solved numerically, but if we were to encounter difficulty then we could use this analysis to provide a more informative boundary condition. These analytical approximations are accurate only for large $x$, but we could use them as guesses for the solver for all $x$. For this purpose it would be natural to impose the boundary conditions $G(0) = 0$ and $F(0) = 0$ to determine $\beta$ and $G(\infty)$.

■ **EXERCISE 3.25**

Experiment with the program `ch3ex5.m` as suggested in Example 3.5.5. In particular, see for yourself how much the various choices for the options affect the run time on your computer. Certainly it is not difficult to generate error messages, but if you have not already done this accidentally then you should try some experiments designed to show how the solver reports failure.

■ **EXERCISE 3.26**

To approximate the solution $U(z)$ of Fisher's BVP, Murray (1993) introduces variables $\xi = z/c$ and $q(\xi) = U(z)$. The ODE becomes

$$c^{-2}q'' + q' + q(1 - q) = 0$$

and the boundary conditions become

$$q(-\infty) = 1, \quad q(\infty) = 0$$

For "large" $c$, the solution $q(\xi)$ is approximated by the outer solution $q_o(\xi)$, that is, the solution of the ODE

$$q_o' + q_o(1 - q_o) = 0$$

Generally such a solution cannot satisfy both boundary conditions, but in this instance the solution

$$q_o(\xi) = \frac{1}{1 + e^\xi}$$

does. Returning to the original independent variable, we have the outer solution

$$U(z) \approx \frac{1}{1 + e^{z/c}}$$

As Murray found by comparing this approximation to numerical solutions, it is a remarkably good approximation. To better understand this, show that the approximation has the correct qualitative behavior near both end points. We found in Section 3.3.2 that (a) as $z \to +\infty$, the solution $U(z) \sim e^{\beta z}$ for $\beta = \left(-c + \sqrt{c^2 - 4}\right)/2$ and (b) as $z \to -\infty$, it satisfies $U'(z) \sim \alpha(U(z) - 1)$ for $\alpha = \left(-c + \sqrt{c^2 + 4}\right)/2$. To get started, note that $q_o \sim e^{-z/c}$ as $z \to +\infty$. Then show that $\beta \approx -1/c$ for "large" $c$. In this, the approximation

$$\sqrt{c^2 - 4} = c\sqrt{1 - 4/c^2} \approx c(1 - 2/c^2)$$

from the binomial series will be needed. Modify the program ch3ex5.m to use this approximate solution (and its derivative) as the guess. Does its use speed up the computation significantly? Plot both the approximation and the numerical solution for some values of $c$ to see how good a guess the approximation is.

■ **EXERCISE 3.27**
Vectorize the evaluation of the ODEs in ch3ex4.m and name the modified program mch3ex4.m. Measure the effect of vectorization on the run time by comparing the result of the *second* invocation

```
>> tic, mch3ex4, toc
```

with the option Vectorized set to on and then set to off. (The second and later invocations have similar run times that may differ significantly from the first invocation.)

■ **EXERCISE 3.28**
Experiment with the BVP of Example 3.5.6. See for yourself how using vectorization and analytical partial derivatives affects the run time on your computer. Modify ch3ex6.m so as to solve the BVP for Reynolds number $R = 1,000,000$. Explore what happens when the solver is not allowed a sufficiently large number of mesh points. Monitor the number of mesh points (using length(sol.x)) that the solver uses to solve the problem for the various values of $R$ on the continuation path.

■ **EXERCISE 3.29**
We solved the BVP of Example 3.5.2 for $\varepsilon = 0.1$, but Keller (1992) is interested in the solution for a range of $\varepsilon$. It is efficient to do this computation by using the solution for one value of $\varepsilon$ as the guess for the next in a continuation process. The ODE is singular

as the parameter $\varepsilon \to 0$ because the order of the equation is reduced, here resulting in an algebraic equation. Typically, singular perturbation problems like this are difficult to solve when $\varepsilon$ is small because there are sharp changes in the solution near one or both end points. This happens because the solution of the ODE for $\varepsilon = 0$ is a solution of a lower-order ODE and thus is not able to satisfy all the boundary conditions simultaneously. The lubrication problem is unusual in that the solution for $\varepsilon = 0$ is

$$y(x) = \sin^2(x), \quad \lambda = 1$$

and it actually satisfies both boundary conditions. Accordingly, there are no boundary layers. If this (outer) solution is used as the guess, then `bvp4c` has no difficulty solving the lubrication problem for small values of $\varepsilon$. However, for insight on how to compute efficiently solutions for several values of a parameter and for practice with continuation, solve the BVP for $\varepsilon = 0.01$ by continuation. Plot the solution and return the value computed for the unknown parameter $\lambda$. Do this by modifying the program `ch3ex2.m` so that you solve the BVP successively for $\varepsilon = \frac{1}{10}, \frac{1}{20}, \ldots, \frac{1}{100}$. The ODE and boundary conditions are simple, so it is easy to provide analytical partial derivatives. Do so and see how much it reduces the run time. The scalar form of the ODE in program `ch3ex2.m` is

```
dydx = (sin(x)^2 - lambda*sin(x)^4/y)/epsilon;
```

Remember that, for the BVP solver, you must vectorize with respect to $x$ as well as $y$. This can be achieved by using

```
dydx = (sin(x) .^2 - lambda*sin(x) .^4 ./ y(1,:))/epsilon;
```

### ■ EXERCISE 3.30

Example 1.10 of Ascher et al. (1995) is a model of the spread of measles that is used to show how to deal with nonseparated boundary conditions. The ODEs

$$y_1' = \mu - \beta(t)y_1y_3$$
$$y_2' = \beta(t)y_1y_3 - \frac{y_2}{\lambda}$$
$$y_3' = \frac{y_2}{\lambda} - \frac{y_3}{\eta}$$

are to be solved on $[0, 1]$ for a periodic solution. That is, the solution vector is to satisfy

$$y(1) = y(0)$$

You are to solve this BVP with `bvp4c`, but if the solver did not provide for nonseparated boundary conditions then you could separate them as suggested by Ascher et al.

(1995): Let $c(t)$ be a vector of three components, and add to the ODEs for $y(t)$ the trivial equations $c' = 0$ that make these components constant. Because they are constant, the periodicity condition can be replaced by the separated boundary conditions $y(0) = c(0)$ and $y(1) = c(1)$.

Solve this BVP for $\mu = 0.02$, $\lambda = 0.0279$, $\eta = 0.01$, and $\beta(t) = \beta_0(1 + \cos 2\pi t)$ where $\beta_0 = 1575$. This BVP can be solved easily with a constant guess, but you might have trouble finding a guess that works. If you are solving this problem in order to gain experience with nonseparated boundary conditions, try something like `1e-3*ones(3,1)`. If you are interested in how you might use continuation to compute a solution of the BVP, read on. In light of Example 3.5.7, it would be natural to approximate the ODEs by their linear part. The trouble with this is that the resulting ODEs do not have a periodic solution! This is easily seen from the linear approximation to the first equation, $y_1' = \mu$. The solution is a straight line with positive slope, so it cannot be periodic. It seems that the nonlinear terms cannot be neglected. On the other hand, we might weaken the nonlinearity so as to obtain a BVP that is easier to solve. A natural way to do this is to regard $\beta_0$ as a parameter. With $\beta_0 = 10$, the nonlinearity is not strong and the BVP is solved easily with a guess of `ones(3,1)`. Using the solution of this BVP as a guess, the BVP with (say) $\beta_0 = 100$ can be solved, and so forth. With the sequence $\beta_0 = 10, 100, 500, 1000$, and 1575, we computed a solution of the given problem without difficulty. Try something along these lines yourself. As in Example 3.5.7, it is interesting to see how the solution changes at each step of continuation, so plot the solution for each $\beta_0$. In plotting the solution, it is convenient to multiply $y_2(t)$ and $y_3(t)$ by 100.

# Chapter 4

## Delay Differential Equations

## 4.1 Introduction

In a system of ordinary differential equations

$$y'(t) = f(t, y(t)) \tag{4.1}$$

the derivative of the solution depends on the solution at the present time $t$. In a system of delay differential equations, the derivative also depends on the solution at earlier times. In this chapter we study DDEs of the form

$$y'(t) = f(t, y(t), y(t - \tau_1), y(t - \tau_2), \ldots, y(t - \tau_k)) \tag{4.2}$$

where the delays (lags) $\tau_j$ are positive constants,

$$0 < \tau_1 < \tau_2 < \cdots < \tau_k$$

It will be convenient to denote the shortest delay by $\tau$ and the longest by $\mathcal{T}$. As illustrated by the survey of Baker, Paul, & Willé (1995a), DDEs arise in models throughout the sciences, but our examples will make clear that they have been especially popular for biological models. DDEs with constant delays are a large and important class. Indeed, Baker and colleagues (1995a) have compiled an extensive bibliography of applications involving DDEs and have pointed out that "[t]he lag functions that arise most frequently in the modelling literature are constants" (Baker, Paul, & Willé 1995b). Furthermore, by restricting attention to problems with constant delays, it is possible to develop software that is more efficient and at the same time more provably reliable than software available for more general problems. Methods used to solve ODEs can generally be extended to solve DDEs. In particular, the MATLAB DDE solver dde23 that we study here is based on the methods used in the MATLAB IVP solver ode23. The user interface of dde23 is much

213

like that of `ode23`, yet owing to differences between DDEs and ODEs, it also resembles the MATLAB BVP solver `bvp4c`.

DDEs and ODEs differ in important ways that are discussed in Section 4.2. In Section 4.3 we explain how the numerical methods developed in Chapter 2 for IVPs can be used to solve DDEs with constant delays. The examples of Section 4.4 show how to solve DDEs with `dde23`; they also highlight differences between DDEs and ODEs. In a final section we describe briefly other kinds of DDEs. They present additional difficulties and so methods and software for solving them are much less developed. Still, there are some useful (Fortran 77) codes for such problems that we discuss briefly.

# 4.2 Delay Differential Equations

We begin with some of the important differences between IVPs for DDEs and ODEs. The most obvious difference is the initial data. The solution of a system of ODEs (4.1) is determined by its value at the initial point $t = a$. In evaluating the DDEs of (4.2), terms like $y(t - \tau_j)$ may represent values of the solution at points prior to the initial point. In particular, when we evaluate the DDEs at the point $t = a$ we must have the value $y(a - \mathcal{T})$. From this we see that, for DDEs, the given initial data must include not only $y(a)$ but also a "history": the values $y(t)$ for all $t$ in the interval $[a - \mathcal{T}, a]$. The numerical solution will be denoted by $S(t)$, so for $t \leq a$ we'll use it to denote the given history.

Because numerical methods for IVPs for both ODEs and DDEs are intended for problems with solutions that have several continuous derivatives, discontinuities in low-order derivatives require special attention. Such discontinuities are not rare for ODEs, but they are almost always present for DDEs because the first derivative of the history function is almost always different from the first derivative of the solution at the initial point. That is, almost always

$$y'(a-) = S'(a-) \neq y'(a+) = f(a, S(a - \tau_1), S(a - \tau_2), \ldots, S(a - \tau_k))$$

There are other ways in which discontinuities in low-order derivatives commonly arise. Some problems have histories with discontinuities in low-order derivatives. For instance, in Exercise 4.8 we consider the solution of an immunology model due to Marchuk. One component of its history for $t \leq 0$ is $\max(0, t + 10^{-6})$, so there is a discontinuity in the first derivative of this component at $t = -10^{-6}$. As with ODEs, a change in the model amounts to a restart and so introduces a discontinuity in the first derivative even when the solution is continuous through the change. This can happen at times known in advance or at times that must be determined by event location. In Exercise 4.7 we consider the solution of a model (due to Hoppensteadt and Waltman) for the spread of an infection. The problem is posed on the interval $[0, 10]$. Because different equations are used to describe different phases of the spread of infection, discontinuities in the first derivative of the solution are introduced at times that are known in advance. Example 4.4.5 features differential equations that change when an event occurs, hence at times that are not known in advance.

Because they propagate, discontinuities are a much more serious matter for DDEs than they are for ODEs. A formal proof of this is clumsy, but it is easy to understand what happens. For a smooth function $f$, the equations (4.2) show that the smoothness of the derivative $y'$ at the current time $t$ depends on the smoothness of the solution $y$ at the past times $t - \tau_j$. Differentiating the equations shows that the same is true for higher derivatives. It will be helpful to have in mind an example that we'll solve analytically in a moment,

$$y'(t) = y(t - 1) \tag{4.3}$$

Obviously $y^{(k+1)}(t) = y^{(k)}(t-1)$ for this equation. In general, if there is a discontinuity at the time $t^*$ of order $k$, meaning that $y^{(k)}$ has a jump at $t = t^*$, then as the variable $t$ moves through $t^* + \tau_j$ there is a discontinuity in $y^{(k+1)}$ because of the term $y(t - \tau_j)$ in equation (4.2). With multiple delays, a discontinuity at the time $t^*$ is propagated to the times

$$t^* + \tau_1, \; t^* + \tau_2, \; \ldots, \; t^* + \tau_k$$

and each of these discontinuities is in turn propagated. If there is a discontinuity at the time $t^*$ of order $k$, then the discontinuity at each of the times $t^* + \tau_j$ is of order at least $k + 1$, and so on. Because the effect of a delay appears in a derivative of higher order, the solution becomes smoother as the integration proceeds. This "smoothing" proves to be quite important to the numerical solution of DDEs. The propagation of discontinuities is taken up in Exercise 4.1. Problems involving discontinuities are solved in Exercises 4.7, 4.8, 4.10, and 4.14.

The *method of steps* is a technique for solving DDEs by reducing them to a sequence of ODEs. To show how it goes and to illustrate the propagation of discontinuities, we solve equation (4.3) with history $S(t) = 1$ for $t \leq 0$. On the interval $0 \leq t \leq 1$, the function $y(t - 1)$ in (4.3) has the known value $S(t - 1) = 1$ because $t - 1 \leq 0$. The DDE on this interval reduces to the ODE $y'(t) = 1$ with initial value $y(0) = S(0) = 1$. We solve this IVP to obtain $y(t) = t + 1$ for $0 \leq t \leq 1$. Notice that the solution of the DDE exhibits a typical discontinuity in its first derivative at $t = 0$ because it is 0 to the left of the origin and 1 to the right. Now that we know the solution for $t \leq 1$, we can reduce the DDE on the interval $1 \leq t \leq 2$ to an ODE $y' = (t - 1) + 1 = t$ with initial value $y(1) = 2$ and solve this IVP to find that $y(t) = 0.5t^2 + 1.5$ on this interval. The first derivative is continuous at $t = 1$, but there is a discontinuity in the second derivative. It is not difficult to see that the DDE's solution on the interval $[k, k + 1]$ is a polynomial of degree $k + 1$ and that the solution has a discontinuity of order $k + 1$ at time $t = k$. We can proceed in exactly the same way with the general equation (4.2). With the history function $S(t)$ defined for $t \le a$, the DDEs reduce to ODEs on the interval $[a, a + \tau]$ because, for each $j$, the argument $t - \tau_j \le t - \tau \le a$ and the $y(t - \tau_j)$ have the known values $S(t - \tau_j)$. Thus, we have an IVP for a system of ODEs with initial value $y(a) = S(a)$. We solve this problem on $[a, a + \tau]$ and extend the definition of $S(t)$ to this interval by taking it to be the solution of this IVP. Now that we know the solution for $t \le a + \tau$, we can move on to the interval $[a + \tau, a + 2\tau]$, and so forth. In this way we can solve the DDEs on the whole interval of interest by solving a sequence of IVPs for ODEs. Although we are mainly interested in problems with constant delays, the method of steps is clearly applicable to DDEs with delays that depend on both $t$ and $y(t)$. The main requirement is simply that the delays all be bounded below by a constant $\tau > 0$.

### ■ EXERCISE 4.1

The following problems test your understanding of the propagation of derivative discontinuities and the method of steps. A computer algebra package such as the Maple kernel of MATLAB or Maple itself would be most helpful in these calculations.

- Solve the DDE

$$y'(t) = [1 + y(t)]y(t - 1)$$

  for $1 \le t \le 3$ with history $y(t) = 1$ for $0 \le t \le 1$. Verify the derivative discontinuities at $t = 1$ and $t = 2$.
- Solve the DDE

$$y'(t) = [1 + y(t)]y(t/2)$$

  for $1 \le t \le 4$ with history $y(t) = 1$ for $\frac{1}{2} \le t \le 1$. Verify the derivative discontinuities at $t = 1$ and $t = 2$. If the solution is continued, at which later times will derivative discontinuities occur?
- Solve the DDE

$$y'(t) = [1 + y^2(t)]y(t - 1)$$

  for $1 \le t \le 2$ with history $y(t) = 1$ for $0 \le t \le 1$. Does the solution extend all the way to $t = 2$?
- Adding a small delay to the effect of a term in an ODE can change the qualitative behavior of solutions. For an example of this, first show that there is a finite point $t^* > 0$ such that the solution of the ODE

$$y'(t) = y^2(t)$$

  with initial value $y(0) = 1$ is not defined for $t \ge t^*$. Then argue that the solution of the DDE

$$y'(t) = y(t)y(t - \tau)$$

with history $y(t) = 1$ for $t \leq 0$ is defined for *all* $t \geq 0$, *no matter how small the delay* $\tau > 0$. To do this, let $y_k(t)$ be the solution of the DDE on $[k\tau, (k+1)\tau]$. First show that $y_0(t)$ exists and is continuous on all of $[0, \tau]$. Next show that, if $y_k(t)$ exists and is continuous on all of $[k\tau, (k+1)\tau]$, then $y_{k+1}(t)$ exists and is continuous on all of $[(k+1)\tau, (k+2)\tau]$. By induction, it then follows that the solution of the DDE exists for all $t \geq 0$.

- All solutions of the ODE

$$y'(t) = -y(t)$$

are decaying exponentials. Show that a delayed effect can lead to a different kind of solution by proving that the DDE

$$y'(t) = -y\left(t - \frac{\pi}{2}\right)$$

has solutions of the form $y(t) = A \sin(t) + B \cos(t)$.

# 4.3 Numerical Methods for DDEs

The method of steps shows that we can solve DDEs with constant delays by solving a sequence of IVPs for ODEs. Because a lot is known about how to solve IVPs, this has been a popular approach to solving DDEs, both analytically and computationally. Solutions smooth out as the integration progresses, so if the shortest delay $\tau$ is small compared to the length of the interval of integration then there can be a good many IVPs, each of which may often be solved in just a few steps. In these circumstances, explicit Runge–Kutta methods are both effective and convenient. Because of this, most solvers are based on explicit Runge–Kutta methods; in particular, the MATLAB DDE solver dde23 is based on the BS(2, 3) pair used by the ODE solver ode23. In what follows we consider how to make the approach practical and use dde23 to illustrate points. More theoretical and practical details for dde23 can be found in Shampine & Thompson (2001).

The example used to explain the method of steps in Section 4.2 will be used here to expose some of the issues. In solving equation (4.3) for the interval $0 \leq t \leq 1$, the DDE reduces to an ODE with $y(t-1)$ equal to the given history $S(t-1)$ and $y(0) = 1$. Solving this IVP with an explicit Runge–Kutta method is perfectly straightforward. A serious complication is revealed when we move to the next interval. The ODE on this interval depends on the solution in the previous interval. However, if we use a Runge–Kutta method in its classical form to compute this solution, we approximate the solution only on a mesh in the interval $[0, 1]$. The first widely available DDE solver, DMRODE (Neves 1975), used cubic Hermite interpolation to obtain the approximate solutions needed at other points in the interval. This approach is not entirely satisfactory because step sizes chosen for an accurate integration may be too large for accurate interpolation. What we need here is a continuous extension of the Runge–Kutta method. The BS(2, 3) Runge–Kutta method used by the code `ode23` was derived along with an accurate continuous extension that happens to be based on cubic Hermite interpolation.

On reaching the current time $t$, we must be able to evaluate the approximate solution $S(t)$ as far back as the point $t - \mathcal{T}$. This means that we must save the information necessary to evaluate the piecewise-polynomial function $S(t)$. The continuous extension of the BS(2, 3) pair is equivalent to cubic Hermite interpolation between mesh points, so it suffices to retain the mesh as well as the value and slope of the approximate solution at each mesh point. The code `dde23` returns the solution as a structure that can have any name, but let us call it `sol`. The mesh is returned in the field `sol.x`. The solution and its slope at the mesh points are returned as `sol.y` and `sol.yp`, respectively. This form of output is an option for `ode23`, but it is the only form of output from `dde23`. Just as with the IVP solvers, the continuous extension is evaluated using the solution structure and the auxiliary function `deval`. It is often useful to be able to evaluate a solution anywhere in the interval of integration, but unlike the situation with the IVP solvers, here we need the capability in order to solve the problem.

Representing the solution as a structure simplifies the user interface. We shall see that, in addition to the information needed for interpolation, we must have other information when solving DDEs – just what depends on the particular problem. Holding this information as fields in a solution structure is both convenient and unobtrusive. All the early DDE solvers were written in versions of Fortran without dynamic storage allocation. This complicates the user interface greatly and there is a real possibility of allocating insufficient storage. The dynamic storage allocation of MATLAB and the use of structures allow a much simpler and more powerful user interface for `dde23`.

The example DDE (4.3) leads to ODEs that are easy to integrate, so a code will try to use large step sizes for the sake of efficiency. Indeed, Runge–Kutta formulas are exact on the first interval, but a solver cannot be permitted to step past the point $t = 1$ because the solution is not smooth there. If the discontinuity is ignored, the order of a Runge–Kutta method can be lowered. The numerical solution is then not as accurate as expected, but what is worse is that the error estimator is not valid. This is because the error is estimated by comparing the results of two formulas and neither has its usual order when the function $f$ is not sufficiently smooth. We can deal with this difficulty by adjusting the step size so that all points where the solution $y(t)$ has a potential low-order discontinuity are mesh points. This implies that none of the functions $y(t), y(t - \tau_1), \ldots, y(t - \tau_k)$ can have a low-order discontinuity in the span of a step from $t_n$ to $t_n + h$. Because we step to discontinuities of the solution $y(t)$, this is clear for $y(t)$ itself. There cannot be a point $\xi$ in $(t_n, t_n + h)$ where some function $y(\xi - \tau_j)$ is not smooth, because the discontinuity in $y(t)$ at the point $\xi - \tau_j$ would have propagated to the point $\xi$ and we would have limited the step size $h$ so that we did not step past this point. Runge–Kutta formulas are one-step formulas and so, if we proceed in this way, they are applied to functions that are smooth in the span of a step and the formulas have the orders expected.

As pointed out earlier, low-order discontinuities are a serious difficulty when solving DDEs because there is almost always one at the initial point and they propagate throughout the interval of integration. On the other hand, the order of a discontinuity increases each time it propagates forward, so we need to track discontinuities only as long as they affect the formulas implemented. Before dde23 begins integrating, it locates all discontinuities of order low enough to affect the integration. It assumes that there will be a discontinuity in the first derivative at the initial point. Some problems have discontinuities at additional points known in advance. To inform dde23 of such derivative discontinuities, the points are provided as the value of the option Jumps. Options are set with the auxiliary function ddeset just as they are set with odeset for the IVP solvers. For instance, the three discontinuities of the Hoppensteadt–Waltman model of Exercise 4.7 can be provided as

```
c = 1/sqrt(2);
options = ddeset('Jumps',[(1-c), 1, (2-c)]);
```

No distinction is made between discontinuities in the history and in the rest of the integration, so the discontinuity of the Marchuk model of Exercise 4.8 is handled in the same way. Exercise 4.10 provides some practice with Jumps. Sometimes the initial value $y(a)$ has a value that is different from the value $S(a)$ of the history. This is handled by supplying $y(a)$ as the value of the option InitialY. When there is a discontinuity in the solution itself at the initial point, we must track it to one level higher than usual. The InitialY option is used in solving the DDE of Exercise 4.6.

Each of the initial discontinuities $\xi$ propagates to the points

$$\xi + \tau_1, \, \xi + \tau_2, \, \ldots, \, \xi + \tau_k$$

where the order of the discontinuity is increased by 1. Each of the resulting discontinuities is, in turn, propagated in the same way. The locations of discontinuities form a tree that we can truncate when the order of the discontinuities is sufficiently high that they do not affect the performance of the formulas implemented. There is a practical difficulty in propagating discontinuities that is revealed by supposing that the DDE has the two lags $\frac{1}{3}$ and 1 and that the integration starts at $t = 0$. The first lag causes discontinuities to appear at the points $0, \frac{1}{3}, 2 \times \frac{1}{3}, 3 \times \frac{1}{3}, \ldots$ and the second causes discontinuities to appear at the points $0, 1, 2, 3, \ldots$. The difficulty is that the finite precision representation of $3 \times \frac{1}{3}$ is not quite equal to 1. It appears to the solver that there are two discontinuities that are extremely close together. This is catastrophic because the step size is limited by the distance between discontinuities. The solver dde23 deals with this by regarding points that differ by no more than ten units of roundoff as being the same and purging one of them.

This purging is done at each level of propagation in order to remove duplicates as early as possible.

The solution of a system of DDEs (4.2) becomes smoother as the integration progresses, which might lead us to expect a corresponding increase in step size. Certainly we must limit the step size so as not to step over a low-order discontinuity, but what happens after they are no longer present? The step size appears to be limited to the shortest delay, for if we were to step from $t_n$ to $t_n + h$ with step size $h > \tau$ then at least one of the arguments $t - \tau_j$ would fall in the interval $(t_n, t_n + h]$. This means that we would need values of the solution at points in the span of the step, but we are trying to compute the solution there and don't yet know these values! Some solvers accept this restriction on the step size. Others, including dde23, use whatever step size appears appropriate to the smoothness of the solution and iterate to evaluate the implicit formulas that arise in this way. On reaching $t_n$ we have a piecewise-cubic polynomial approximation $S(t)$ to the solution for $t \leq t_n$. When the BS(2, 3) formulas need values $y(t - \tau_j)$ for arguments $t - \tau_j > t_n$, these values are predicted by extrapolating the polynomial approximation of the preceding interval. After evaluating the formulas we have a new cubic polynomial approximation for the solution on the interval $(t_n, t_n + h]$, and we use it when correcting the solution by reevaluating the formulas. Evaluating the BS(2, 3) formula when the step size is larger than the shortest delay is quite like evaluating an implicit multistep method.

Earlier we mentioned the need for event location. This capability is available in dde23 exactly as in ode23 except that information about events is always returned as fields of the solution structure. When a terminal event is located, it is not unusual to continue the integration after modifying the equations and possibly modifying the final value of the solution for use as the initial value of the new integration. This is easy enough when solving ODEs with the IVP solvers of MATLAB because the solutions of the various IVPs can be aggregated easily to obtain a solution over the whole range of interest. The situation is quite different when solving DDEs. The most important difference is that a history must be supplied for the subsequent integration. This history is mainly the solution as computed up to the event, which is to be evaluated by interpolation, but it may also include the given history, which may be supplied in three different forms in dde23 and so be evaluated in different ways. The solver dde23 accepts a solution structure as history and uses information stored in this structure to evaluate properly the terms in the DDE that involve delays. Another issue is the propagation of discontinuities. It may be that the event occurs whilst some of the propagated discontinuities are still active. For the current integration we must reconstruct the tree of discontinuities and propagate them into the current interval of integration. This requires some information to be retained from the previous integration. There is, of course, a new discontinuity introduced at the new initial point. It is not unusual for the initial value of the solution to be different from the last solution value of the previous integration. This is handled with the InitialY option. When a solution structure is provided as history, dde23 incorporates it into the solution structure that is returned. In this way, the solution structure returned by the solver always provides the solution on the whole interval of integration.

# 4.4 Solving DDEs in MATLAB

Because of the more general nature of DDEs, it is necessary to provide more information to DDE solvers than to ODE solvers. Although the manner in which this information is provided varies between solvers, the same basic information is required for all. Here we see how to solve DDEs with dde23. It is limited to problems with constant delays, but the examples and exercises of this section show that, for this class of problems, it is both easy to use and powerful. Solving a DDE with dde23 is much like solving an ODE with ode23, but there are some notable differences. Some of our examples have been viewed as a considerable challenge for DDE solvers in general scientific computing.

## EXAMPLE 4.4.1

In their paper "An Epidemic Model with Recruitment–Death Demographics and Discrete Delays", Genik & van den Driessche (1999) consider the total population $N(t)$ to be divided into four states: $S(t)$, susceptible; $E(t)$, exposed but not infective; $I(t)$, infective; and $R(t)$, recovered. So, the total population is

$$N(t) = S(t) + E(t) + I(t) + R(t)$$

The five parameters of the model and values for the *Pasteurella muris* virus in laboratory mice are: the birth rate, $A = 0.330$; the natural death rate, $d = 0.006$; the contact rate of infective individuals, $\lambda = 0.308$; the rate of recovery, $\gamma = 0.040$; and the excess death rate for infective individuals, $\varepsilon = 0.060$. There are two time delays in the model: a temporary immunity delay, $\tau = 42.0$, and a latency delay (the time before becoming infective after exposure), $\omega = 0.15$. The DDEs of the model are

$$\frac{dS(t)}{dt} = A - dS(t) - \lambda \frac{S(t)I(t)}{N(t)} + \gamma I(t - \tau)e^{-d\tau}$$

$$\frac{dE(t)}{dt} = \lambda \frac{S(t)I(t)}{N(t)} - \lambda \frac{S(t - \omega)I(t - \omega)}{N(t - \omega)}e^{-d\omega} - dE(t)$$

$$\frac{dI(t)}{dt} = \lambda \frac{S(t - \omega)I(t - \omega)}{N(t - \omega)}e^{-d\omega} - (\gamma + \varepsilon + d)I(t)$$

$$\frac{dR(t)}{dt} = \gamma I(t) - \gamma I(t - \tau)e^{-d\tau} - dR(t)$$

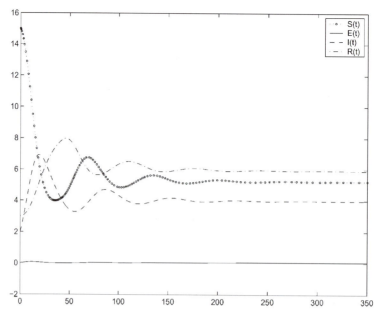

Figure 4.1: The SEIRS epidemic model of Genik & van den Driessche (1999).

The equations are to be solved on [0, 350] with history defined by $S(t) = 15$, $E(t) = 0$, $I(t) = 2$, and $R(t) = 3$ for all times $t \le 0$. Some analytical work reveals the existence of a stability threshold $R_0 = \lambda e^{-d\omega}/(\gamma + \varepsilon + d)$. The disease dies out if $R_0 < 1$ but does not die out if $R_0 > 1$. For the preceding choice of parameters, $R_0 \approx 2.9$.

The straightforward solution of a DDE with `dde23` resembles so closely solving an IVP with `ode23` that we'll just give a program and explain the differences. The output is displayed as Figure 4.1.

```
function sol = ch4ex1
global tau omega
tau  = 42.0; omega = 0.15;

sol = dde23(@ddes, [tau, omega], [15; 0; 2; 3], [0, 350]);
plot(sol.x,sol.y)
legend('S(t)', 'E(t)', 'I(t)', 'R(t)')

%=========================================
function dydt = ddes(t,y,Z)
global tau omega
```

```
% Parameters:
A = 0.330; d = 0.006; lambda = 0.308;
gamma = 0.040; epsilon = 0.060;

% Variable names used in stating the DDEs:
S = y(1); E = y(2); I = y(3); R = y(4);
% Z(:,1) corresponds to the lag tau.
Itau    = Z(3,1);
% Z(:,2) corresponds to the lag omega.
Somega = Z(1,2); Eomega = Z(2,2);
Iomega = Z(3,2); Romega = Z(4,2);

Noft = S + E + I + R;
Nomega = Somega + Eomega + Iomega + Romega;
dSdt = A - d*S - lambda*((S*I)/Noft) + gamma*Itau*exp(-d*tau);
dEdt = lambda*((S*I)/Noft) - ...
       lambda*((Somega*Iomega)/Nomega)* exp(-d*omega) - d*E;
dIdt = lambda*((Somega*Iomega)/Nomega)*exp(-d*omega) ...
       - (gamma+epsilon+d)*I;
dRdt = gamma*I - gamma*Itau*exp(-d*tau) - d*R;

dydt = [ dSdt; dEdt; dIdt; dRdt];
```

A call to `dde23` has the form `sol = dde23(@ddes,lags,history,tspan)`. Just as when calling one of the MATLAB IVP solvers, `tspan` is the interval of integration, but there are some small differences in the way this array is used. If you specify more than two entries in `tspan`, the IVP solvers return values of the solutions at these points; `dde23` determines output at specific points in a different manner, so only the first and last entries of `tspan` are meaningful to this solver. Further, for `tspan` equal to $[t_0, t_f]$, `dde23` requires that $t_0 < t_f$. The `history` argument can have three forms. One is a handle for a function that evaluates the solution at given $t \leq t_0$ and returns it as a column vector. Here this function might have been coded as

```
function v = history(t)
v = [15; 0; 2; 3];
```

The history is constant for this example. This is so common that the solver also allows the history to be supplied in the form of a constant vector, and that is what is done in `ch4ex1.m`. (The third form of the history function is a solution structure; it is used only when restarting an integration, which is not done in this example.) The delays are provided as a vector `lags`, here `[tau, omega]`. Just as when using a function `odes` to evaluate ODEs, `ddes` is a function that evaluates the DDEs. Also as with ODEs, the input argument `t` is the current value of $t$ and the input argument `y` is an approximation to the solution $y(t)$. What is different when solving DDEs is an input array `Z`. It contains approximations to the solution at all the delayed arguments. Specifically, `Z(:,j)` approximates the function $y(t - \tau_j)$ for the lag $\tau_j$ given as `lags(j)`. As with ODEs, the function `ddes` must return a column vector.

The output of `dde23` is a solution structure, here called `sol`. The mesh is returned in the field `sol.x` and the solution and its slope at these mesh points are returned as `sol.y` and `sol.yp`, respectively. As mentioned earlier, this form of output is an option for `ode23`, but it is the only form of output from `dde23`. You can invoke `ch4ex1.m` with no output arguments, but if you wish to retain and study the solution then you should invoke it as

```
>> sol = ch4ex1;
```

You might, for example, wish to compute and plot the total population $N(t)$ on return from the program. This can be done at the command line with

```
>> Noft = sum(sol.y);
>> plot(sol.x,Noft)
```

## EXAMPLE 4.4.2

We show how to obtain output at specific points with Example 5 of Willé & Baker (1992), a scalar equation that exhibits chaotic behavior. We solve the DDE

$$y'(t) = \frac{2y(t-2)}{1 + y(t-2)^{9.65}} - y(t) \tag{4.4}$$

on the interval of integration [0, 100] with history $y(t) = 0.5$ for $t \le 0$.

Here `dde23` computes an approximate solution $S(t)$ valid throughout the range `tspan` and places in the structure `sol` the information necessary to evaluate it with `deval`. All you must do is supply the solution structure and an array `t` of points where you want to evaluate the solution,

```
S = deval(sol,t);
```

With this form of output, you can solve a DDE just once and then obtain inexpensively as many solution values as you like, anywhere you like. The numerical solution is itself continuous, so you can always plot a smooth graph by evaluating it at sufficiently many points using `deval`.

Willé & Baker (1992) plot the function $y(t - 2)$ against the function $y(t)$. This is quite a common task in nonlinear dynamics, but we cannot proceed as in Example 4.4.1. This is

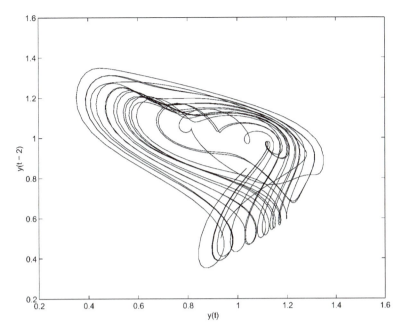

Figure 4.2:  Example 5 of Willé & Baker (1992).

because the entries of `sol.x` are not equally spaced: If the point $\hat{t}$ appears in `sol.x` then we have an approximation to $y(\hat{t})$ in `sol.y`, but generally the point $\hat{t} - 2$ does not appear in `sol.x` and so we do not have an approximation to $y(\hat{t} - 2)$. The function `deval` makes such plots easy. In `ch4ex2.m` we first define an array `t` of 1000 equally spaced points in the interval $[2, 100]$ and obtain solution values at these points using `deval`. We then use the function `deval` a second time to evaluate the solution at the entries of an array of values $t - 2$. In this way we obtain values approximating both the function $y(t)$ and the function $y(t - 2)$ for the same values of $t$. This might seem like a lot of plot points, but `deval` is just evaluating a piecewise-cubic polynomial function and is coded to take advantage of fast built-in functions and vectorization, so this is not expensive and results in the smooth graph of Figure 4.2.

The complete program to compute and plot the function $y(t - 2)$ against the function $y(t)$ is

```
function sol = ch4ex2

sol = dde23(@ddes,2,0.5,[0, 100]);

t = linspace(2,100,1000);
y = deval(sol,t);
```

```
ylag = deval(sol,t - 2);
plot(y,ylag)
xlabel('y(t)');
ylabel('y(t - 2)');


%===================================
function dydt = ddes(t,y,Z)
dydt = 2*Z/(1 + Z^9.65) - y;
```

## EXAMPLE 4.4.3

We consider a model (Corwin, Sarafyan, & Thompson 1997) of multiple long-term part-
nerships and HIV transmission in a homogeneous population. Formulating this model as
a set of ODEs and DDEs is somewhat complicated, but once we have the equations it is
easy enough to solve them with dde23. If you wish, you can proceed directly to the equations and the discussion of how to solve them. In the model, $x$ is the number of susceptible
individuals, $y$ is the number of infected individuals, $\lambda$ is the rate at which susceptible individuals become infected per unit time, $D$ is the duration of a long-term partnership, $G$ is
the rate of recovery from an infective stage back to the susceptible stage, $c$ is the number
of sexual contacts with long-term partners leading to infection per person per unit time,
and $n$ is the size of the initial population. For times $t \leq D$ the equations are

$$x'(t) = -x(t)\lambda(t) + Gy(t)$$
$$n = x(t) + y(t)$$
$$\lambda(t) = \frac{c}{n}\left\{\int_0^t \int_s^t e^{-G(t-w)}x(w)\lambda(w)\,dw\,ds + \int_0^t e^{-G(t-s)}y(s)\,ds\right\}$$

and for times $D \leq t$ they are

$$x'(t) = -x(t)\lambda(t) + Gy(t)$$
$$n = x(t) + y(t)$$
$$\lambda(t) = \frac{c}{n}\left\{\int_{t-D}^t \int_s^t e^{-G(t-w)}x(w)\lambda(w)\,dw\,ds + \int_{t-D}^t e^{-G(t-s)}y(s)\,ds\right\}$$

This is a set of Volterra integro-differential equations. Sometimes such equations can
be solved using techniques for DDEs, and that is the case here. Indeed, for $t \leq D$ we can
proceed by solving the ODEs

$$x'(t) = -x(t)\lambda(t) + Gy(t)$$

$$y'(t) = -x'(t)$$

$$\lambda'(t) = (c/n)e^{-Gt}\{(I_1'(t) + I_2'(t)) - G(I_1(t) + I_2(t))\}$$

$$I_1'(t) = e^{Gt}y(t)$$

$$I_2'(t) = te^{Gt}x(t)\lambda(t)$$

$$I_3'(t) = e^{Gt}x(t)\lambda(t)$$

and for $D \le t$, by solving the DDEs

$$x'(t) = -x(t)\lambda(t) + Gy(t)$$

$$y'(t) = -x'(t)$$

$$\lambda'(t) = (c/n)e^{-Gt}\{(I_1'(t) + I_2'(t)) - G(I_1(t) + I_2(t))\}$$

$$I_1'(t) = e^{Gt}y(t) - e^{G(t-D)}y(t - D)$$

$$I_2'(t) = De^{Gt}x(t)\lambda(t) - I_3(t)$$

$$I_3'(t) = e^{Gt}x(t)\lambda(t) - e^{G(t-D)}x(t - D)\lambda(t - D)$$

We don't need an ODE for the component $y(t) = n - x(t)$, but it is convenient to introduce one so that we obtain values for this important quantity along with $x(t)$ and $\lambda(t)$. To see how we have dealt with the integrals in $\lambda(t)$, let's work through the more complicated case of $D \le t$. First we write the function $\lambda(t)$ as

$$\lambda(t) = \frac{c}{n}e^{-Gt}(I_1(t) + I_2(t)) \tag{4.5}$$

where

$$I_1(t) = \int_{t-D}^{t} e^{Gs}y(s)\, ds$$

$$I_2(t) = \int_{t-D}^{t} \int_{s}^{t} e^{Gw}x(w)\lambda(w)\, dw\, ds$$

Differentiating the expression (4.5) gives the stated ODE for the function $\lambda(t)$, so we just need to compute the derivatives of the integrals. Using Leibnitz's rule, the first is easy:

$$I_1'(t) = e^{Gt}y(t) - e^{G(t-D)}y(t - D)$$

The second is a more complicated differentiation that leads to

$$I_2'(t) = \int_{t-D}^{t} e^{Gt}x(t)\lambda(t)\, ds - \int_{t-D}^{t} e^{Gs}x(s)\lambda(s)\, ds$$

The first integral on the right is trivial, and if we define

$$I_3(t) = \int_{t-D}^{t} e^{Gs} x(s) \lambda(s) \, ds$$

then by differentiating we can compute the second by solving the DDE

$$I_3'(t) = e^{Gt} x(t) \lambda(t) - e^{G(t-D)} x(t-D) \lambda(t-D)$$

Finally,

$$I_2'(t) = De^{Gt} x(t) \lambda(t) - I_3(t)$$

and we have verified all the DDEs for this case.

The derivation of the equations for $t \leq D$ is similar, but there appears to be no need for the integral $I_1(t)$ in this case. We do need it to deal with values for the integrals when we switch to solving DDEs at the time $t = D$. By definition, the integrals all vanish at $t = 0$, providing initial values for their ODEs. Comparing the definitions in the two cases shows that the integrals are continuous at $t = D$, providing the initial values we need for integration of the DDEs. A convenient way to deal with $I_1(t)$ then is to introduce the equation stated for $t \leq D$. It is defined so that we integrate it to obtain the value we need at $t = D$ for the second integration.

For a concrete example, we solve this problem when the constants have values

$$c = 0.5, \quad n = 100, \quad G = 1, \quad D = 5$$

and the initial values are

$$x(0) = 0.8n, \quad y(0) = 0.2n, \quad \lambda(0) = 0$$

In coding the equations we can, of course, rewrite the equation for $y'(t)$ by using the equation for $x'(t)$, so that there is no derivative on the right-hand side. In principle this must be done, but the equations stated show the model and derivation more clearly and serve to illustrate a useful technique for coding the evaluation of the equations: instead of eliminating $x'(t)$ in the equation for $y'(t)$, just evaluate $x'(t)$ first and use it in evaluating $y'(t)$. Similarly, $I_1'(t)$ and $I_2'(t)$ would be evaluated and then used to evaluate $\lambda'(t)$.

The solver dde23 makes no assumption that the functions $y(t - \tau_j)$ actually appear in the equations. Because of this, you can use it to solve ODEs. It is expecting to solve a DDE, so you must include Z as an input argument to the function defining the equations. If you solve an ODE, it is best to input an empty array for lags because otherwise the solver will track potential discontinuities and restrict the step size accordingly even though there are no discontinuities due to delays. The program ch4ex3.m does this, first solving the ODEs for $t \leq D$ and then solving the DDEs for $D \leq t$. The most important point illustrated by this example is that the solution structure returned in one integration

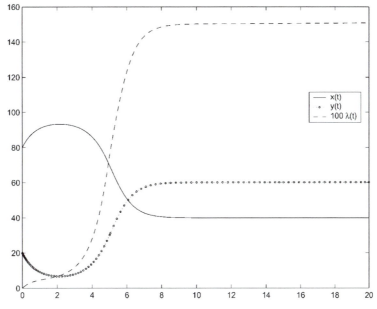

Figure 4.3: HIV multiple partnership problem.

can be used directly as history for another, a matter discussed at length in Section 4.3. Furthermore, after each integration, the solution structure is extended so as to provide the solution for the whole integration. This makes it easy to plot $x$, $y$, and $100\lambda$ as Figure 4.3.

```
function sol = ch4ex3
global c D G n
c = 0.5; D = 5; G = 1; n = 100;

% x = y(1), y = y(2), lambda = y(3),
% I_1 = y(4), I_2 = y(5), I_3 = y(6).
y0 = [0.8*n; 0.2*n; 0; 0; 0; 0];

sol = dde23(@odes,[],y0,[0, D]);
sol = dde23(@ddes,D,sol,[D,4*D]);

plot(sol.x,[sol.y(1:2,:); 100*sol.y(3,:)]);
legend('x(t)','y(t)','100\lambda(t)',0)

%==============================================================
function dydt = odes(t,y,Z)
global c D G n
```

```
dydt = zeros(6,1);
dydt(1) = - y(1)*y(3) + G*y(2);
dydt(2) = - dydt(1);
dydt(4) = exp(G*t)*y(2);
dydt(5) = t*exp(G*t)*y(1)*y(3);
dydt(6) = exp(G*t)*y(1)*y(3);
dydt(3) = (c/n)*exp(-G*t)*((dydt(4)+dydt(5))-G*(y(4)+y(5)));

function dydt = ddes(t,y,Z)
global c D G n
dydt = zeros(6,1);
dydt(1) = - y(1)*y(3) + G*y(2);
dydt(2) = - dydt(1);
dydt(4) = exp(G*t)*y(2) - exp(G*(t - D))*Z(2);
dydt(5) = D*exp(G*t)*y(1)*y(3) - y(6);
dydt(6) = exp(G*t)*y(1)*y(3) - exp(G*(t - D))*Z(1)*Z(3);
dydt(3) = (c/n)*exp(-G*t)*((dydt(4)+dydt(5))-G*(y(4)+y(5)));
```

Early attempts to solve this problem treated it as an IVP for $x(t)$ and $y(t)$ with a complicated coefficient $\lambda(t)$. Quadrature routines were used to approximate the integrals defining this coefficient, but the matter is difficult because the integrands involve $x$, $y$, and $\lambda$ itself. As a consequence, the first program for this problem that we encountered ran for hours on a mainframe computer and often crashed. Improvements in the analytical and numerical treatment of the task have reduced the computation time to less than a second on a PC. The problem is considered further in Exercise 4.5.

### EXAMPLE 4.4.4

A model of the infamous four-year life cycle of a population of lemmings is found in Tavernini (1996). The equation

$$y'(t) = ry(t)\left(1 - \frac{y(t - 0.74)}{m}\right) \tag{4.6}$$

is solved on $[0, 40]$. The parameters have values $r = 3.5$ and $m = 19$. Notice that, with these values, the equation has a constant (steady-state) solution of $y(t) = m = 19$. Tavernini uses this solution as history and perturbs the initial value so that the solution $y(t)$ will move away from the steady state. Here we use the initial value $y(0) = 19.001$ for this purpose. When the initial value is different from the history, all you must do is provide it as the value of the InitialY option. The jump at the initial point is so small that we

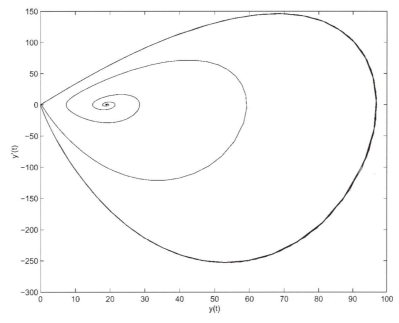

Figure 4.4: Population of lemmings; phase plane.

must specify tolerances more stringent than the defaults so that the solver will "notice" it. These tolerances are specified exactly as for IVPs but using ddeset instead of odeset.

Because the solution settles into a periodic behavior, it is instructive to plot the derivative $y'(t)$ against the solution $y(t)$. This is easily done because, in addition to sol.y approximating $y(t)$ at the mesh points, there is a field sol.yp approximating $y'(t)$. Figure 4.4 shows the phase-plane plot produced by ch4ex4.m.

The minima and maxima of the population are of some interest, so we use their computation to illustrate event location. Not much need be said because this is done exactly as with the IVP solvers, but the task has points of interest. Local minima are found as points where $y'(t) = 0$, but of course this equation is satisfied for maxima as well. The two kinds of events are distinguished by a local minimum occurring where the derivative $y'(t)$ increases through zero and a local maximum where it decreases. The direction argument of the events function makes this distinction. The solver dde23 returns the same information about events as ode23, but as fields in the solution structure instead of optional output arguments. In ch4ex4.m we use the find command to determine the indices in the output array that correspond to the two different kinds of events. With these indices, we can extract the information for the kind of event that interests us. In particular, the plot of the solution $y$ against time in Figure 4.5 shows the local minima as filled squares and the local maxima as open circles.

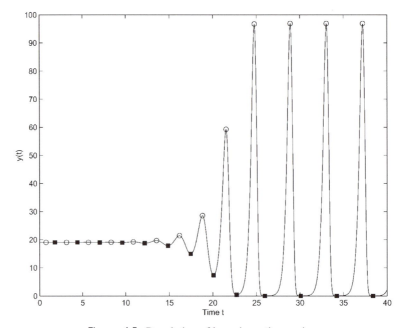

Figure 4.5: Population of lemmings; time series.

The problem is solved and the output plotted in a straightforward way by ch4ex4.m. We have left the constants $r$ and $m$ as parameters and, for variety, we have passed them through the call list of dde23 as optional arguments. As with the IVP solvers, they must then appear as arguments to the functions for evaluating the DDEs and events, even if they are not used by the function.

```
function sol = ch4ex4
r = 3.5; m = 19;
options = ddeset('Events',@events,'InitialY',19.001,...
                 'RelTol',1e-4,'AbsTol',1e-7);
sol = dde23(@ddes,0.74,19,[0, 40],options,r,m);
plot(sol.y,sol.yp);
xlabel('y(t)');
ylabel('y''(t)');

n1 = find(sol.ie == 1);
x1 = sol.xe(n1);
y1 = sol.ye(1,n1);
n2 = find(sol.ie == 2);
x2 = sol.xe(n2);
y2 = sol.ye(1,n2);
```

```
figure
plot(sol.x,sol.y,'k',x1,y1,'rs',x2,y2,'bo')
xlabel('Time t');
ylabel('y(t)');


%===========================================================
function dydt = ddes(t,y,Z,r,m)
dydt = r*y*(1 - Z/m);

function [value,isterminal,direction] = events(t,y,Z,r,m)
dydt = ddes(t,y,Z,r,m);
value = [dydt; dydt];
direction = [+1; -1];
isterminal = [0; 0];
```

## EXAMPLE 4.4.5

We have seen examples and exercises with discontinuities at times known in advance that have been handled with the `Jumps` and `InitialY` options. Other discontinuities depend on the solution and so must be located using the `Events` option. Events that lead to changes in the equations must be followed by a restart. We have already seen an example of restarting at a time known in advance. However, when events depend on the solution, we do not know where and how many events will occur, so we do not know in advance where and how many restarts there will be. The example we take up now has this complication. Setting it up and solving it is a challenge to the user interface of any DDE solver. Another example of the use of the `Events` option is found in Exercise 4.14.

A two-wheeled suitcase may begin to rock from side to side as it is pulled. When this happens, the person pulling it attempts to return it to the vertical by applying a restoring moment to the handle. There is a delay in this response that can significantly affect the stability of the motion. This is modeled by Suherman et al. (1997) with the DDE

$$\theta''(t) + \text{sign}(\theta(t))\gamma \cos(\theta(t)) - \sin(\theta(t)) + \beta\theta(t - \tau) = A \sin(\Omega t + \eta)$$

where $\theta(t)$ is the angle of the suitcase to the vertical. This equation is solved on the interval of integration [0, 12] as a pair of first-order equations with $y_1(t) = \theta(t)$ and $y_2(t) = \theta'(t)$. Figure 3 of Suherman et al. (1997) shows the solution component $y_1(t)$ plotted against time $t$ and the phase-plane plot of $y_2(t)$ plotted against $y_1(t)$ when

$$\gamma = 2.48, \quad \beta = 1, \quad \tau = 0.1, \quad A = 0.75, \quad \Omega = 1.37, \quad \eta = \arcsin(\gamma/A)$$

and the initial history is the constant vector zero. A wheel hits the ground (the suitcase is vertical) when $y_1(t) = 0$. The integration is then to be restarted with $y_1(t) = 0$ and

$y_2(t)$ multiplied by the coefficient of restitution, here chosen to be 0.913. The suitcase is considered to have fallen over when $|y_1(t)| = \pi/2$ and the run is then terminated.

This problem is solved with

```
function sol = ch4ex5a
state = +1;
opts = ddeset('RelTol',1e-5,'Events',@events);
sol = dde23(@ddes,0.1,[0; 0],[0 12],opts,state);

ref = [4.516757065, 9.751053145, 11.670393497];
fprintf('Kind of Event:                      dde23    reference\n');
event = 0;
while sol.x(end) < 12
    event = event + 1;
    if sol.ie(end) == 1
        fprintf('A wheel hit the ground.   %10.4f   %10.6f\n',...
                sol.x(end),ref(event));
        state = - state;
        opts = ddeset(opts,'InitialY',[ 0; 0.913*sol.y(2,end)]);
        sol = dde23(@ddes,0.1,sol,[sol.x(end) 12],opts,state);
    else
        fprintf('The suitcase fell over.   %10.4f   %10.6f\n',...
                sol.x(end),ref(event));
        break;
    end
end
plot(sol.y(1,:),sol.y(2,:))
xlabel('\theta(t)')
ylabel('\theta''(t)')

%==================================================================
function dydt = ddes(t,y,Z,state)
gamma = 0.248; beta  = 1; A = 0.75; omega = 1.37;
ylag = Z(1,1); dydt = [y(2); 0];
dydt(2) = sin(y(1)) - state*gamma*cos(y(1)) - beta*ylag ...
            + A*sin(omega*t + asin(gamma/A));

function [value,isterminal,direction] = events(t,y,Z,state)
value = [y(1); abs(y(1))-pi/2];
isterminal = [1; 1]; direction = [-state; 0];
```

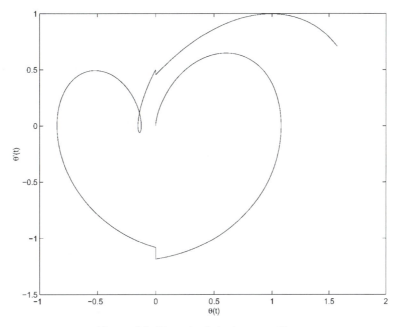

Figure 4.6: Two-wheeled suitcase problem.

The program produces the phase-plane plot of Figure 4.6, which agrees with that of Suherman et al. (1997). It also reports what kind of event occurred and the location of the event:

```
>> ch4ex5a;
Kind of Event:                    dde23    reference
A wheel hit the ground.          4.5168     4.516757
A wheel hit the ground.          9.7511     9.751053
The suitcase fell over.         11.6704    11.670393
```

The reference values were computed with the DKLAG5 code used in Suherman et al. (1997) and verified with its successor DKLAG6, which is described briefly in Section 4.5.

This is a relatively complicated model, so we will elaborate on some aspects of the program. Coding of the DDE is straightforward except for evaluating properly the discontinuous coefficient $\text{sign}(y_1(t))$. This is accomplished by initializing a parameter state to $+1$ and changing its sign whenever dde23 returns because $y_1(t)$ vanished. Handling state in this manner ensures that dde23 does not need to deal with the discontinuities it would otherwise see if the derivative were coded in a manner that allowed state to change before the integration is restarted; see Shampine & Thompson (2000) for a discussion of this issue. After a call to dde23, we must consider why it has returned. One possibility is that it has reached the end of the interval of integration – as indicated by the last point reached, `sol.x(end)`, being equal to 12. Another is that the suitcase has fallen over, as indicated by `sol.ie(end)` being equal to 2. Both cases cause termination of the run. More interesting is a return because a wheel hit the ground, $y_1(t) = 0$, which is indicated by `sol.ie(end)` being equal to 1. The sign of `state` is then changed and the integration restarted. Because the wheel bounces, the solution at the end of the current integration, `sol.y(:,end)`, must be modified for use as initial value of the next integration. The `InitialY` option is used to deal with an initial value that is different from the history. The event $y_1(t) = 0$ that terminates one integration occurs at the initial point of the next integration. As with the IVP solvers, the solver `dde23` does not terminate the run in this special situation of an event at the initial point. No special action is necessary, but the solver does locate and report an event at the initial point, so it is better practice to avoid this by defining more carefully the event function. When the indicator `state` is +1 (resp., −1), we are interested in locating where the solution component $y_1(t)$ vanishes only if it decreases (resp., increases) through zero. We inform the solver of this by setting the first component of the argument `direction` to `-state`. Notice that `ddeset` is used to alter an existing options structure in the `while` loop. This is a convenient capability also present in `odeset`, the corresponding function for IVPs. The rest of the program is just a matter of reporting the results of the computations. Default tolerances give an acceptable solution, though the phase-plane plot would benefit from plotting more solution values. Reducing the relative error tolerance to `1e-5` gives better agreement with the reference values.

This computation is fast enough that we can go on to discuss other computations of interest that require it as an auxiliary computation. For a fixed set of parameters, the critical excitation amplitude $A_{cr}$ is the smallest value of $A$ for which the suitcase overturns. Suherman et al. (1997) give graphs of $A_{cr}$ as a function of the excitation frequency $\Omega$ for several sets of problem parameters. For the set $\beta = 1$, $\tau = 0.1$, and $\Omega = 2$, the critical amplitude is between 0.4 and 1.4. We can compute $A_{cr}$ by defining a function $f(A)$ to have value +1 if the suitcase does not fall over in the course of the integration and value −1 if it does. We then use bisection to find where the function $f(A)$ changes sign. In this approach we assume that the suitcase does not fall over for $A < A_{cr}$ and does fall over for larger values of $A$. (An interesting aspect of this problem is that, for some choices of the parameters, the second assumption is not always valid.) For the interval of integration we use $\left[0, \frac{40\pi}{\Omega}\right]$, corresponding to twenty cycles of the excitation moment. The function is evaluated by a version of the program `ch4ex5a.m` with its output removed and the excitation amplitude $A$ passed as a parameter. The change of sign is located with the MATLAB function `fzero`. To keep down the run time, we use modest tolerances both in the root-finder `fzero` and in the solver `dde23`. After all, each evaluation of the function $f(A)$ in `fzero` requires the solution of a DDE. To give us something to look at whilst the program is running, we trace the progress of fzero. The critical amplitude is found to be about $A_{cr} = 0.93$. The program is

```
function ch4ex5b
options = optimset('Display','iter','TolX',0.01);
Acr = fzero(@f,[0.4, 1.4],options);
fprintf('\nThe critical excitation amplitude is %4.2f.\n',Acr);


%===================================================
function fval = f(A)
fval = +1;
omega = 2;
tfinal = 40*pi/omega;
state = +1;
opts = ddeset('Events',@events);
sol = dde23(@ddes,0.1,[0; 0],[0 tfinal],opts,state,A);
while sol.x(end) < tfinal
   if sol.ie(end) == 1
      state = - state;
      opts = ddeset(opts,'InitialY',[ 0; 0.913*sol.y(2,end)]);
      sol = dde23(@ddes,0.1,sol,[sol.x(end) tfinal],opts,state,A);
   else
      fval = -1;
      break;
   end
end


function dydt = ddes(t,y,Z,state,A)
omega = 2; gamma = 0.248; beta = 1;
ylag = Z(1,1);
dydt = [y(2); 0];
dydt(2) = sin(y(1)) - state*gamma*cos(y(1)) - beta*ylag ...
          + A*sin(omega*t + asin(gamma/A));

function [value,isterminal,direction] = events(t,y,Z,state,A)
value = [y(1); abs(y(1))-pi/2];
isterminal = [1; 1];
direction = [-state; 0];
```

### ■ EXERCISE 4.2

An epidemic model due to Cooke (see MacDonald 1978) uses the following equation to describe the fraction $y(t)$ at time $t$ of a population that is infected:

$$y'(t) = by(t - 7)[1 - y(t)] - cy(t)$$

Here $b$ and $c$ are positive constants. The equation is solved on the interval $[0, 60]$ with history $y(t) = \alpha$ for $t \leq 0$. The constant $\alpha$ satisfies $0 < \alpha < 1$.

Write a function with input arguments $\alpha$, $b$, and $c$ to solve this DDE with `dde23` and plot the solution. The title of the plot should give the values of $\alpha$, $b$, and $c$ that were used. For all values of $b$ and $c$, $y(t) = 0$ is obviously an equilibrium point (steady-state solution). For $b > c$, the solution $y(t) = 1 - c/b$ is a second equilibrium point. If you experiment with values for $\alpha$, $b$, and $c$, you will find that when $b > c$, the solution approaches the second equilibrium point and otherwise it approaches the first. As a specific example, compute and plot the approach of the solution to the nontrivial equilibrium solution when $\alpha = 0.8$, $b = 2$, and $c = 1$.

### ■ EXERCISE 4.3

A problem with a history that is not constant is solved by Neves (1975). In this problem the DDEs

$$y_1'(t) = y_5(t - 1) + y_3(t - 1)$$
$$y_2'(t) = y_1(t - 1) + y_2(t - 0.5)$$
$$y_3'(t) = y_3(t - 1) + y_1(t - 0.5)$$
$$y_4'(t) = y_5(t - 1)y_4(t - 1)$$
$$y_5'(t) = y_1(t - 1)$$

are to be integrated for $0 \leq t \leq 1$ with history

$$y_1(t) = e^{t+1}$$
$$y_2(t) = e^{t+0.5}$$
$$y_3(t) = \sin(t + 1)$$
$$y_4(t) = y_1(t)$$
$$y_5(t) = y_1(t)$$

for $t \leq 0$. Solve this problem and plot all components of the solution. Your program will be much like `ch4ex1.m`, but you must evaluate the history in a (sub)function and supply its handle as the history argument of `dde23`. Remember that the functions for evaluating the DDEs and the history must return column vectors.

■ **EXERCISE 4.4**

Farmer (1982) gives plots of various Poincaré sections for the Mackey–Glass equation, a scalar DDE that exhibits chaotic behavior. Reproduce Figure 2a of Farmer (1982) by solving the DDE

$$y'(t) = \frac{0.2y(t-14)}{1 + y(t-14)^{10}} - 0.1y(t)$$

on [0, 300] with history $y(t) = 0.5$ for $t \leq 0$ and plotting the function $y(t-14)$ against the function $y(t)$. The figure begins at $t = 50$ in order to allow an initial transient sufficient time to settle down. To reproduce the figure, form an array of 1000 equally spaced points in the interval [50, 300], evaluate the function $y(t)$ at these points, and then evaluate the function $y(t-14)$. Your program will be much like `ch4ex2.m`.

■ **EXERCISE 4.5**

The solution of the HIV multiple partnership problem computed in Example 4.4.3 with `ch4ex3.m` appears to approach a steady state: a constant solution $x_s$, $y_s$, $\lambda_s$. Show that there are two steady-state solutions – namely,

$$\lambda_s = 0, \quad x_s = n, \quad y_s = 0$$

and the interesting one,

$$\lambda_s = cD - G, \quad x_s = \frac{Gn}{\lambda_s + G}, \quad y_s = n - x_s$$

To do this, go back to the integral form of the problem for $D \leq t$ and assume that the solution is constant for large $t$. Modify `ch4ex3.m` to solve the model for values $G = 0.1, 1$, and 2 in turn and verify that the limit values `sol.y(1:3,end)` are in reasonable agreement with the analytical steady-state solution. In this model, the constant $G$ is the rate of recovery from an infective stage back to the susceptible stage.

■ **EXERCISE 4.6**

The manual (Paul 1995) for the Fortran 77 code `ARCHI` discussed in Section 4.5 provides a sample program for solving the DDEs

$$y_1'(t) = y_1(t-1)y_2(t-2)$$
$$y_2'(t) = -y_1(t)y_2(t-2)$$

on the interval [0, 4] with history $y_1(t) = \cos(t)$ and $y_2(t) = \sin(t)$ for $t < 0$ and initial values $y_1(0) = 0$ and $y_2(0) = 0$. Notice that $y_1(t)$ is discontinuous at the initial point, so the option `InitialY` must be used to supply the solution there. For practice, compute and plot the solution. The sample program specifies a pure absolute error tolerance of $10^{-9}$. The code `dde23` does not permit a pure absolute error, but for practice with options, use the default relative error tolerance and set `AbsTol` to `1e-9`.

■ **EXERCISE 4.7**

Example 4.4 of Oberle & Pesch (1981) is an infection model due to Hoppensteadt and Waltman. The equation

$$y'(t) = \begin{cases} -ry(t)0.4(1-t), & 0 \le t \le 1-c \\ -ry(t)(0.4(1-t)+10-e^{\mu}y(t)), & 1-c < t \le 1 \\ -ry(t)(10-e^{\mu}y(t)), & 1 < t \le 2-c \\ -re^{\mu}y(t)(y(t-1)-y(t)), & 2-c < t \end{cases}$$

is solved on the interval of integration $[0, 10]$ with history $y(t) = 10$ for $t \le 0$. Here $c = 1/\sqrt{2}$ and $\mu = r/10$. Oberle & Pesch (1981) solve this problem for several values of the parameter $r$, so in your code make $r$ a parameter but solve the problem just for $r = 0.5$, a case for which Oberle and Pesch provide the reference value $y(10) = 0.06302089869$. The different phases of the spread of the disease are described by different equations. The model requires the solution to be continuous, but the changes in the equation defining $y'(t)$ lead to jumps in the low-order derivatives. Because this happens at times that are known in advance, all you must do is provide the solver with these times as the value of the `Jumps` option. You can code the DDE in a straightforward way by using an `if` construct. Because the reference solution was computed with much more stringent tolerances, use a relative error tolerance of `1e-5` and an absolute error tolerance of `1e-8`. An interesting aspect of this problem is that, in addition to the solution $y(t)$, an approximation to the function

$$I(t) = -\frac{y'(t)}{ry(t)}$$

is required. Using `sol.yp` and `sol.y`, plot this function.

■ **EXERCISE 4.8**

The equations of the Marchuk immunology model discussed in Hairer, Nörsett, & Wanner (1987) are

$$y_1'(t) = (h_1 - h_2 y_3(t))y_1(t)$$
$$y_2'(t) = \xi(y_4(t))h_3 y_3(t-0.5)y_1(t-0.5) - h_5(y_2(t)-1)$$
$$y_3'(t) = h_4(y_2(t) - y_3(t)) - h_8 y_3(t)y_1(t)$$
$$y_4'(t) = h_6 y_1(t) - h_7 y_4(t)$$

Here the coefficient

$$\xi(y_4(t)) = \begin{cases} 1 & \text{if } y_4(t) \leq 0.1 \\ \frac{10}{9}(1 - y_4(t)) & \text{if } 0.1 < y_4(t) \leq 1 \end{cases}$$

is continuous, but it has a jump in its first derivative where the solution component $y_4(t) = 0.1$, which leads to a jump in a low-order derivative of the solution component $y_2(t)$. The problem is solved on $[0, 60]$ with history

$$y_1(t) = \max(0, t + 10^{-6}), \quad y_2(t) = 1, \quad y_3(t) = 1, \quad y_4(t) = 0$$

for $t \leq 0$. As noted previously, the solution component $y_1(t)$ has a jump in its first derivative at the point $t = -10^{-6}$ that propagates into the interval of integration. Figure 15.8 of Hairer et al. (1987) presents plots for parameter values

$$h_1 = 2, \quad h_2 = 0.8, \quad h_3 = 10^4, \quad h_4 = 0.17, \quad h_5 = 0.5, \quad h_7 = 0.12, \quad h_8 = 8$$

and for two values of $h_6$, namely $h_6 = 10$ and $h_6 = 300$. Use dde23 to solve the problem for $h_6 = 300$. To reproduce the plot of Hairer et al. (1987), you will need to scale the components as

$$10^4 y_1, \ 0.5 y_2, \ y_3, \ 10 y_4$$

and use axis([0 60 -1 15.5]). An array yplot of scaled values for plotting can be formed easily with

```
yplot = sol.y;
yplot(1,:) = 1e4*yplot(1,:);
```

and similar commands for the other components. To solve this problem accurately over the whole interval of integration, you will need to reduce the tolerances to (say) a relative tolerance of 1e-5 and an absolute tolerance of 1e-8. Use the Jumps option to tell the solver about the discontinuity at the point $t = -10^{-6}$. Terminate the integration when the event function $y_4(t) - 0.1$ vanishes. Use a parameter state with value $+1$ if $y_4(t) \leq 0.1$ and $-1$ otherwise. The problem is to be solved with $y_4(0) = 0$, so initialize state to $+1$. Thereafter, each time that the solver returns, check whether you have reached the end of the interval of integration. If sol.x(end) < 60, change the sign of state and call the solver dde23 again with the previous solution as history. In the function for evaluating the DDEs, set $\xi(y_4(t)) = 1$ if state is $+1$ and $\xi(y_4(t)) = \frac{10}{9}[1 - y_4(t)]$ otherwise. You'll need a history function, so remember that if you pass state as an optional argument to dde23 then you must also make it an input argument of the history function.

### ■ EXERCISE 4.9

Hale (1971) cites predator–prey models obtained by (a) introducing a resource limitation on the prey and (b) assuming the birth rate of predators responds to changes in the magnitude of the population $y_1$ of prey and the population $y_2$ of predators only after a time delay $\tau$. Starting with the system of ODEs

$$y_1'(t) = ay_1(t) + by_1(t)y_2(t)$$
$$y_2'(t) = cy_2(t) + dy_1(t)y_2(t)$$

(see Ortega & Poole 1981), we arrive in this way at a system of DDEs

$$y_1'(t) = ay_1(t)\left(1 - \frac{y_1(t)}{m}\right) + by_1(t)y_2(t)$$
$$y_2'(t) = cy_2(t) + dy_1(t - \tau)y_2(t - \tau)$$

It is interesting to explore the effect of the delay, so solve both systems on the interval $[0, 100]$ with initial values $y_1(0) = 80$ and $y_2(0) = 30$ for the ODEs and the same values as constant history for the DDEs. Suppose that the parameters are

$$a = 0.25, \quad b = -0.01, \quad c = -1.00, \quad d = 0.01, \quad m = 200$$

Recall that you solve ODEs with `dde23` by setting `lags` to `[]`. You must write the function for evaluating the differential equations to accept an input argument `Z`. When the array of lags is empty, `dde23` calls this function with an empty array for `Z`. You can use this fact to code the evaluation of both sets of equations in the same function by testing `isempty(Z)` to find out which set to evaluate. A more straightforward approach is to use different functions for the ODEs and the DDEs. Solve the DDEs with $\tau = 1$. Plot in one figure the component $y_2(t)$ against the component $y_1(t)$ for both the ODEs and DDEs. This phase-plane plot of the solution of the ODEs should be a closed curve corresponding to a limit cycle. To achieve this you will need to tighten the error tolerances. For example, with a command like

```
options = ddeset('RelTol',1e-5,'AbsTol',1e-8);
```

you should obtain a plot like Figure 4.7. You might experiment with the tolerances to see how small you need to set them in order to compute a closed curve in the phase plane.

The figure makes clear that introducing a delay into an ODE model can have a profound effect on the behavior of the solution. By experimenting with $\tau$ you will find this to be true even for small delays. It is also interesting to remove the resource term $1 - y_1(t)/m$ and then see how the orbits change as $\tau$ is changed.

### ■ EXERCISE 4.10

A cardiovascular model due to Ottesen (1997) involves the arterial pressure, $P_a(t) = y_1(t)$, the venous pressure, $P_v(t) = y_2(t)$, and the heart rate, $H(t) = y_3(t)$. Ottesen studies

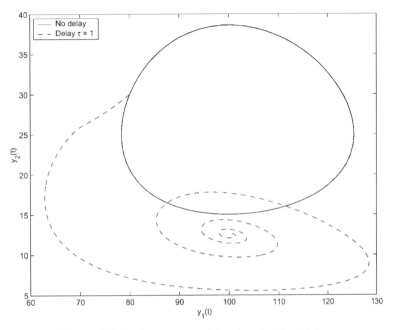

Figure 4.7: Predator–prey model with and without delay.

conditions under which the delay causes qualitative differences in the solution and, in particular, oscillations in $P_a(t)$. Delays $\tau = 1.0$, 1.4, 3.9, 5.0, 7.5, and 10 are considered in Ottesen (1997). Compute and plot the arterial pressure for $\tau$ equal to 1.0 and 7.5, values leading to solutions that differ dramatically. Solve on the interval [0, 350] the equations

$$y_1'(t) = -\frac{1}{c_a R} y_1(t) + \frac{1}{c_a R} y_2(t) + \frac{1}{c_a} V_{str} y_3(t)$$

$$y_2'(t) = \frac{1}{c_v R} y_1(t) - \left(\frac{1}{c_v R} + \frac{1}{c_v r}\right) y_2(t)$$

$$y_3'(t) = f(T_s, T_p)$$

where

$$T_s = \left[1 + \left(\frac{y_1(t - \tau)}{\alpha_s}\right)^{\beta_s}\right]^{-1}$$

$$T_p = \left[1 + \left(\frac{\alpha_p}{y_1(t)}\right)^{\beta_p}\right]^{-1}$$

$$f(T_s, T_p) = \frac{\alpha_H T_s}{1 + \gamma_H T_p} - \beta_H T_p$$

For $t \leq 0$, the solution has the constant value

$$y_1(t) = P_0$$

$$y_2(t) = \left(\frac{r}{r+R}\right) P_0$$

$$y_3(t) = \left(\frac{1}{r+R}\right)\left(\frac{P_0}{V_{str}}\right)$$

As in Ottesen (1997), use

$$c_a = 1.55, \quad c_v = 519, \quad R = 1.05, \quad r = 0.068, \quad \alpha_0 = \alpha_s = \alpha_p = 93, \quad \alpha_H = 0.84$$

and

$$\beta_0 = \beta_s = \beta_p = 7, \quad \beta_H = 1.17, \quad \gamma_H = 0, \quad V_{str} = 67.9, \quad P_0 = 93$$

One of the figures of Ottesen (1997) shows the solution components when, beginning at $t = 600$, the peripheral pressure $R$ is reduced exponentially from its constant value of $R = 1.05$ to a constant value of $R = 0.84$. The change in $R$ leads to a sharp change in the heart rate. For this computation, the delay was $\tau = 4$ and the interval $[0, 1000]$. Modify your program so that it solves this problem and plots the heart rate. All you must do is (a) inform the solver of the low-order discontinuity at a known time by setting the value of the option Jumps to 600 and (b) modify the function for evaluating the DDEs to include

```
if t <= 600
    R = 1.05;
else
    R = 0.21 * exp(600-t) + 0.84;
end
```

### ■ EXERCISE 4.11

Plant's neuron interaction model (see MacDonald 1989) is given by the equations

$$y_1'(t) = y_1(t) - \frac{y_1^3(t)}{3} - y_2(t) + m(y_1(t-\tau) - y_{1,0})$$

$$y_2'(t) = r(y_1(t) + a - by_2(t))$$

Here $y_{1,0}$ is the first component of a steady-state solution $(y_{1,0}, y_{2,0})$. With the parameters set to

$$a = 0.8, \quad b = 0.7, \quad r = 0.08$$

solve the equations on $[0, 60]$ with history

$$y_1(t) = 0.4y_{1,0}, \quad y_2(t) = 1.8y_{2,0}$$

Plant chose the unique real steady state $y_{1,0}$ for which $y_{1,0}^2 > 1 - rb$. After a little analytical work, you can use the MATLAB function `roots` to find that $y_{1,0} = -1.22764016121492$. Use this value of $y_{1,0}$ to compute the value $y_{2,0}$. Solve the DDEs for $\tau = 20$ and separately $m = +10$ and $m = -10$. Plot the solutions. You will find that they can exhibit very sharp changes.

■ **EXERCISE 4.12**

A population growth model due to Cooke, van den Driessche, & Zou (1999) considers the effects of a delay in maturation and a nonlinear birth rate. It describes the population $y(t)$ by the DDE

$$y'(t) = be^{-ay(t-T)}y(t-T)e^{-d_1 T} - dy(t)$$

The DDE is to be solved on the interval $[0, 25]$ with history $y(t) = 3.5$ for $t \leq 0$. Generally problems are solved for a number of choices of parameters. For practice, solve this problem for each of the data sets

1.  $a = 1, d = 1, d_1 = 1, b = 20$
2.  $a = 1, d = 1, d_1 = 1, b = 80$
3.  $a = 1, d = 1, d_1 = 0, b = 20$
4.  $a = 1, d = 1, d_1 = 0, b = 80$

You can do this in four runs if you like, but you could do it in a single run by defining an array such as B = [20 80 20 80] and solving the DDE in a `for` loop with index dataset and data such as b = B(dataset). For each data set, solve the DDE using three values of the delay, $T = 0.2$, 1.0, and 2.4, and then plot the solutions in the same figure. You will find that the size of the delay profoundly affects the behavior of the solution. To make the point that solution structures can be indexed, we note that if you define an array Delays = [0.2 1.0 2.4] then you could code the three computations as

```
for i = 1:3
    T = Delays(i);
    sol(i) = dde23(@ddes,T,3.5,[0, 25],opts);
end
```

On exit from this loop, the mesh and solution for the first delay is sol(1).x and sol(1).y, and similarly for the other delays. The value of $T$ must be communicated to the subfunction ddes as a parameter or a global variable because it appears in the DDE. In this code fragment it was communicated as a global variable along with the parameters of the data set. In your computations use ddeset to specify tolerances more stringent than the default values, namely RelTol = 1e-5 and AbsTol = 1e-8.

### ■ EXERCISE 4.13

Martin & Ruan (2001) consider the effects of delay and constant prey harvesting on the solution of several common predator–prey models. In this exercise you are to study numerically one or more of the models they investigate. In all cases, $x(t)$ is the number of predators at time $t$ and $y(t)$ is the number of prey.

- Martin and Ruan first consider models in which a delay appears in the term governing the growth of prey. In this they present analytical and numerical results for the system

$$x'(t) = x(t)\left[2\left(1 - \frac{x(t-\tau)}{40}\right) - \frac{y(t)}{x(t)+10}\right] - 10$$

$$y'(t) = y(t)\left[\frac{x(t)}{x(t)+10} - \frac{2}{3}\right]$$

Show that $(20, 15)$ is an equilibrium point for this system. For $\tau = 0$, solve the ODEs on $[0, 100]$ with initial values $x(0) = 40$ and $y(0) = 16$. (Use dde23 with an empty array for lags.) Plot $(x(t), y(t))$ to verify numerically that the equilibrium point is asymptotically stable for this value of $\tau$. For $\tau = 0.826$, solve the DDEs on $[0, 100]$ with constant history $x = 40$ and $y = 2$. Plot $(x(t), y(t))$ to verify numerically that there is a limit cycle about the equilibrium point for this value of $\tau$. Reduce the default RelTol to $10^{-5}$ in both integrations.

- The second kind of model considered by Martin and Ruan contains delays in the predator response function. Results are given for the system

$$x'(t) = x(t)\left[2\left(1 - \frac{x(t)}{50}\right) - \frac{y(t)}{x(t)+40}\right] - 10$$

$$y'(t) = y(t)\left[-3 + \frac{6x(t-\tau)}{x(t-\tau)+40}\right]$$

Show that $(40, 12)$ is an equilibrium point for this system. For each of $\tau = 7$ and $\tau = 9$, solve the DDEs on $[0, 250]$ with constant history $x = 44$ and $y = 2$. Plot $(x(t), y(t))$ to verify numerically that the equilibrium point is asymptotically stable for $\tau = 7$ and that there is a limit cycle about the equilibrium point for $\tau = 9$. Reduce the default RelTol to $10^{-5}$ in both integrations.

- Finally, Martin and Ruan study the effects of adding constant prey harvesting to a well-known model. They present results for the system

$$x'(t) = x(t)[20 - x(t) - y(t)] - 7$$

$$y'(t) = -15y(t) + 3x(t-\tau)y(t-\tau)$$

Show that $\left(5, \frac{68}{5}\right)$ is an equilibrium point for this system. For $\tau = 0.05$, solve the DDEs on $[0, 10]$ with constant history $x = 2$ and $y = 10$. Plot $(x(t), y(t))$ to verify numerically that there is a limit cycle about the equilibrium point. Reduce the default `RelTol` to $10^{-5}$.

### ■ EXERCISE 4.14

The controller problem of Marriott & DeLisle (1989) is a DDE that involves a step function of the delayed solution. With $\Delta = y(t - 12) - x_b$, the equation is

$$y'(t) = \tau^{-1}\left(-y(t) + \pi\left(a + \varepsilon\,\text{sign}(\Delta) - u\sin^2(\Delta)\right)\right)$$

It is solved on the interval $[0, 120]$ with history $y(t) = 0.6$ for $t \leq 0$ and parameter values

$$x_b = -0.427, \quad a = 0.16, \quad \varepsilon = 0.02, \quad u = 0.5, \quad \tau = 1$$

The code `dde23` is sufficiently robust that it can solve this problem without any special provision for $\text{sign}(\Delta)$. However, ignoring discontinuities can get you into trouble even when solving an ODE (Shampine & Thompson 2000), much less a DDE. You can be more confident of your numerical results if you arrange for the solver always to be integrating an equation with smooth coefficients. This can be done much as in program `ch4ex5a.m` by letting the parameter `state` be the value the solver is to use for $\text{sign}(\Delta)$. With the given history, this parameter is initialized to $+1$. Use the `Events` option to terminate the integration if $y(t - 12) - x_b = 0$. If this terminal event occurs before you reach the end of the interval, change the sign of `state` and start a new integration on the interval `[sol.x(end), 120]`. The current solution structure is the history structure for the new integration. Print out a message and the starting point of each new integration.

# 4.5 Other Kinds of DDEs and Software

So far we've considered equations of the form

$$y'(t) = f(t, y(t), y(t - \tau_1), y(t - \tau_2), \ldots, y(t - \tau_k)) \tag{4.7}$$

with constant delays $\tau_j$. The simplest extension is to allow the delays to depend on the independent variable $t$. Provided that the delays are bounded away from zero, such problems can be solved much like problems with constant delays. It is more difficult and much more expensive to work out how discontinuities propagate. If there is a low-order discontinuity at $t^*$ then it is "felt" in equation (4.7) when some $t - \tau_j(t) = t^*$. That is, for each delay the discontinuity propagates to the first zero of the function

$$t - \tau_j(t) - t^* = 0 \tag{4.8}$$

When the delay is constant, the solution of this equation is trivial and we find again that the discontinuity propagates to the point $t^* + \tau_j$. When the delay is not constant, the zero must be found numerically. Sometimes this is difficult, even in principle. For instance, it is difficult to recognize the presence of a zero of even multiplicity, and the location of any zero of multiplicity greater than 1 is ill-conditioned. If a delay can vanish, new conceptual and computational problems arise. For example, the DDE

$$y'(t) = y\left(\frac{t}{2}\right) = y\left(t - \frac{t}{2}\right)$$

for $t \geq 0$ makes no use of values prior to the initial point $t = 0$; in other words, this DDE does not require a history function. For the sake of efficiency, dde23 resorts to implicit formulas and to iteration when dealing with delays that are smaller than the natural step size. This is indispensable for delays that vanish or tend to zero. Robust codes for finding zeros can deal with functions that are not smooth, but they are much more efficient when the functions are smooth. An equation used to illustrate both kinds of difficulties is

$$y'(t) = y(t)(1 - y([t]))$$

Here, the greatest integer function $[t] = n$ for $n \leq t < n + 1$, so the delay $\tau(t) = t - [t]$ is discontinuous and vanishes at the integers.

Delays that depend on the solution itself are said to be state-dependent. They change everything. When the delays depend only on time, the propagation equation (4.8) can be solved before beginning the integration, just as with constant delays. Now the equation is

$$t - \tau_j(t, y(t)) - t^* = 0$$

and we need $y(t)$ to find out how delays propagate so that we can compute $y(t)$. Obviously this implicit relationship requires some kind of iteration. As a practical matter, it must be resolved at each step. Some codes predict a solution over the span of a tentative step by extrapolating the current solution. Using this predicted solution, they check for discontinuities in the span of the tentative step, reduce the step size as necessary, and iterate. This procedure is expensive, and some aspects of current implementations are less than completely satisfactory. An example with a known solution is

$$y_1'(t) = y_2(t)$$
$$y_2'(t) = y_2(e^{1-y_2(t)})y_2^2(t)e^{1-y_2(t)}$$

to be solved on $[0.1, 5]$ with history taken from the analytical solution $y_1(t) = \ln(t)$ and $y_2(t) = t^{-1}$. Notice that, in addition to the complications of a state-dependent delay $t - e^{(1-y_2(t))}$, the delay vanishes at $t = 1$.

Table 4.1: *Statistics for* DKLAG6 *solving a neutral DDE with vanishing delay.*

| Tol | Steps | Evals | ERO1 | ERO2 | Ratio |
|---|---|---|---|---|---|
| $10^{-2}$ | 2 | 50 | 0.191e−02 | 0.191e−02 | 1.00 |
| $10^{-4}$ | 3 | 77 | 0.298e−01 | 0.464e−01 | 1.55 |
| $10^{-6}$ | 6 | 212 | 0.283e+00 | 0.308e+00 | 1.09 |
| $10^{-8}$ | 11 | 347 | 0.524e−01 | 0.187e+00 | 3.57 |
| $10^{-10}$ | 21 | 554 | 0.242e+00 | 0.157e+01 | 6.48 |

So far, we've considered equations for which the derivative of the solution at the current time $t$ depends on the solution then and the solution at one or more previous times. It may also depend on the *derivative* of the solution at previous times. Such equations are said to be of *neutral* type. An example with a known solution is

$$y'(t) = \cos(t)[1 + y(ty^2(t))] + 0.3y(t)y'(ty^2(t))$$
$$+ 0.7\sin(t)\cos(t\sin^2(t)) - \sin(t + t\sin^2(t)) \tag{4.9}$$

to be solved on the interval $\left[0, \frac{\pi}{2}\right]$ with the history taken from the analytical solution $y(t) = \sin(t)$. Solving DDEs of neutral type is a real challenge and the subject of current research. One difficulty is that the existence, uniqueness, and continuous dependence of a solution on the data of the problem are not clear. Unlike the other DDEs we have discussed, discontinuities in derivatives are not smoothed as the integration proceeds. This presents obvious difficulties for efficient numerical integration of the equations. Clearly we must retain an approximation to the derivative of the solution. Unfortunately, when a continuous extension produces an approximation $S(t)$ to the solution $y(t)$ of a certain order of accuracy, usually the approximation $S'(t)$ to $y'(t)$ has one lower order of accuracy.

Despite the difficulties we have pointed out and others, the DDE solvers discussed next contain provisions for solving neutral problems and have good track records. To make the point, we solved equation (4.9) using DKLAG6. In addition to the fundamental difficulty of the derivative at the current time depending on the derivative at a previous time, this equation is difficult because the delay depends on both $t$ and $y$ and vanishes at both the initial and final times. In Table 4.1 we present some statistics from this computation. Similar results are obtained when using any of the cited solvers. In the table:

- Tol is the error tolerance;
- Steps is the number of integration steps;
- Evals is the number of derivative evaluations;

- ERO1 is the ratio of the maximum error to the error tolerance at any integration mesh point;
- ERO2 is the ratio of the maximum error to the error tolerance at any point in the interval of integration; and
- Ratio is the ratio of the maximum global interpolation error and the maximum global integration error.

Although solving the kinds of problems described in this section is still the subject of research, some effective solvers are widely available. We have already mentioned ARCHI (Paul 1995) and DKLAG6 (Corwin et al. 1997; Corwin & Thompson 1996); another solver familiar to us is DDVERK (Enright & Hayashi 1997, 1998). All three solvers accept neutral problems and problems with state-dependent delays. All three allow small and vanishing delays. Each is based on an explicit Runge–Kutta pair with continuous extension and is coded in Fortran 77.

ARCHI is based on a $(4, 5)$ pair. A matter worth some discussion is that discontinuity tracking is optional in this solver. The local error control of explicit Runge–Kutta methods is sufficiently robust that generally it can recognize, locate, and step to discontinuities on its own. Relying on local error control is attractive because it asks much less of a user, but the approach is less reliable and possibly less accurate than tracking the discontinuities. ARCHI allows the user to specify either extrapolation or iterative evaluation of implicit formulas.

DDVERK is based on a $(5, 6)$ pair. Small and vanishing delays are handled iteratively. Discontinuities are detected by means of a defect error control. Suspected discontinuities are located and special interpolants are used when stepping over a discontinuity.

DKLAG6 is also based on a $(5, 6)$ pair. As with ARCHI, discontinuity tracking is optional. DKLAG6 is the only one of the solvers that provides for event location. The user interface is radically different from the other solvers – virtually all communication is through user-provided subroutines. A tool (see Corwin & Thompson 1993) is available for generating these subroutines and for plotting solutions computed by DKLAG6.

# Bibliography

W. Ames & E. Lohner (1981). Nonlinear models of reaction–diffusion in rivers. In R. Vichnevetsky & R. Stepleman (Eds.), *Advances in Computer Methods for Partial Differential Equations,* vol. IV, pp. 217–19. New Brunswick, NJ: IMACS.

P. Amodio, J. R. Cash, G. Roussos, R. W. Wright, G. Fairweather, I. Gladwell, G. L. Kraut, & M. Paprzycki (2000). Almost block diagonal linear systems: Sequential and parallel solution techniques, and applications. *Numer. Lin. Alg. Appl.* 7: 275–317.

D. Arnold & J. C. Polking (1999). *Ordinary Differential Equations Using MATLAB,* 2nd ed. Englewood Cliffs, NJ: Prentice-Hall.

U. M. Ascher, J. Christiansen, & R. D. Russell (1979). COLSYS – A collocation code for boundary value problems. In B. Childs et al. (Eds.), *Codes for Boundary Value Problems* (Lecture Notes in Comput. Sci., 76), pp. 164–85. New York: Springer-Verlag.

U. M. Ascher, J. Christiansen, & R. D. Russell (1981). Collocation software for boundary value ODE's. *ACM Trans. Math. Software* 7: 209–29.

U. M. Ascher, R. M. M. Mattheij, & R. D. Russell (1995). *Numerical Solution of Boundary Value Problems for Ordinary Differential Equations.* Philadelphia: SIAM.

U. M. Ascher & R. D. Russell (1981). Reformulation of boundary value problems into "standard" form. *SIAM Review* 23: 238–54.

G. Bader & U. Ascher (1987). A new basis implementation for a mixed order boundary value solver. *SIAM J. Sci. Stat. Comput.* 9: 483–500.

P. B. Bailey, B. S. Garbow, H. G. Kaper, & A. Zettl (1991). Eigenvalue and eigenfunction computations for Sturm–Liouville problems. *ACM Trans. Math. Software* 17: 491–9.

P. B. Bailey, M. K. Gordon, & L. F. Shampine (1978). Automatic solution of the Sturm–Liouville problem. *ACM Trans. Math. Software* 4: 193–208.

P. B. Bailey, L. F. Shampine, & P. E. Waltman (1968). *Nonlinear Two Point Boundary Value Problems.* New York: Academic Press.

C. T. H. Baker, C. A. H. Paul, & D. R. Willé (1995a). A bibliography on the numerical solution of delay differential equations. Numerical Analysis Report no. 269, Mathematics Department, University of Manchester, U.K.

C. T. H. Baker, C. A. H. Paul, & D. R. Willé (1995b). Issues in the numerical solution of evolutionary delay differential equations. *Adv. Comput. Math.* 3: 171–96.

C. M. Bender & S. A. Orszag (1999). *Advanced Mathematical Methods for Scientists and Engineers I, Asymptotic Methods and Perturbation Theory.* New York: Springer-Verlag.

P. Bogacki & L. F. Shampine (1989). A 3(2) pair of Runge–Kutta formulas. *Appl. Math. Lett.* 2: 1–9.

R. L. Borrelli & C. S. Coleman (1999). *ODE Architect.* New York: Wiley.

R. W. Brankin, J. R. Dormand, I. Gladwell, P. Prince, & W. L. Seward (1989). ALGORITHM 670: A Runge–Kutta–Nyström code. *ACM Trans. Math. Software* 15: 31–40.

R. W. Brankin & I. Gladwell (1994). A Fortran 90 version of RKSUITE: An ODE initial value solver. *Ann. Numer. Math.* 1: 363–75.

R. W. Brankin, I. Gladwell, & L. F. Shampine (1993). RKSUITE: A suite of explicit Rünge–Kutta codes. In R. P. Agarwal (Ed.), *Contributions to Numerical Mathematics* (WSSIAA, 2), pp. 41–53. Singapore: World Scientific.

K. E. Brenan, S. L. Campbell, & L. R. Petzold (1996). *Numerical Solution of Initial-Value Problems in Differential-Algebraic Equations* (SIAM Classics in Applied Mathematics, 14). Philadelphia: SIAM.

P. N. Brown, G. D. Byrne, & A. C. Hindmarsh (1989). VODE: A variable coefficient ODE solver. *SIAM J. Sci. Stat. Comput.* 10: 1038–51.

J. R. Cash & M. H. Wright (1991). A deferred correction method for nonlinear two-point boundary value problems: Implementation and numerical evaluation. *SIAM J. Sci. Stat. Comput.* 12: 971–89.

T. K. Caughy (1970). Large amplitude whirling of an elastic string – A nonlinear eigenvalue problem. *SIAM J. Appl. Math.* 18: 210–37.

T. Cebeci & H. B. Keller (1971). Shooting and parallel shooting methods for solving the Falkner–Skan boundary-layer equations. *J. Comp. Phys.* 7: 289–300.

J. D. Cole (1968). *Perturbation Methods in Applied Mathematics.* Waltham, MA: Blaisdell.

K. Cooke, P. van den Driessche, & X. Zou (1999). Interaction of maturation delay and nonlinear birth in population and epidemic models. *J. Math. Biol.* 39: 332–52.

S. P. Corwin, D. Sarafyan, & S. Thompson (1997). DKLAG6: A code based on continuously imbedded sixth order Runge–Kutta methods for the solution of state dependent functional differential equations. *Appl. Numer. Math.* 24: 319–33.

S. P. Corwin & S. Thompson (1993). DRAKE: Continuous simulation software for the solution of delay differential equations on personal computers. Computer Science Department Technical Report Series, no. TR-93-001, Radford University, Radford, VA.

S. P. Corwin & S. Thompson (1996). DKLAG6: Solution of systems of functional differential equations with state dependent delays. Computer Science Department Technical Report Series, no. TR-96-002, Radford University, Radford, VA.

A. R. Curtis, M. J. D. Powell, & J. K. Reid (1974). On the estimation of sparse Jacobian matrices. *J. Inst. Math. Appl.* 13: 117–19.

H. T. Davis (1962). *Introduction to Nonlinear Differential and Integral Equations.* New York: Dover.

F. R. de Hoog & R. Weiss (1976). Difference methods for boundary value problems with a singularity of the first kind. *SIAM J. Numer. Anal.* 13: 775–813.

F. R. de Hoog & R. Weiss (1978). Collocation methods for singular boundary value problems. *SIAM J. Numer. Anal.* 15: 198–217.

J. R. Dormand (1996). *Numerical Methods for Differential Equations.* Boca Raton, FL: CRC Press.

J. R. Dormand & P. J. Prince (1980). A family of embedded Runge–Kutta formulae. *J. Comput. Appl. Math.* 27: 19–26.

C. H. Edwards (1997). Newton's nose-cone problem. *Mathematica J.* 7: 64–71.

W. H. Enright & H. Hayashi (1997). A delay differential equation solver based on a continuous Runge–Kutta method with defect control. *Numer. Algorithms* 16: 349–64.

W. H. Enright & H. Hayashi (1998). Convergence analysis of the solution of retarded and neutral differential equations by continuous methods. *SIAM J. Numer. Anal.* 35: 572–85.

W. H. Enright & P. H. Muir (1996). Runge–Kutta software with defect control for boundary value ODEs. *SIAM J. Sci. Comput.* 17: 479–97.

J. D. Farmer (1982). Chaotic attractors of an infinite-dimensional dynamical system. *Physica D* 4: 366–93.

E. Fehlberg (1970). Klassiche Runge-Kutta-Formeln vierter und niedrigerer Ordnung mit Schrittenweiten-Kontrolle und ihre Anwendung auf Wärmeleitungsprobleme. *Computing* 6: 61–71.

B. A. Finlayson (1972). *The Method of Weighted Residuals and Variational Principles.* New York: Academic Press.

C. A. J. Fletcher (1983). *Computational Galerkin Methods.* New York: Springer-Verlag.

*GAMS.* The *Guide to Available Mathematical Software* is available at ⟨http://gams.nist.gov/⟩.

C. W. Gear (1971). *Numerical Initial Value Problems in Ordinary Differential Equations.* Englewood Cliffs, NJ: Prentice-Hall.

L. Genik & P. van den Driessche (1999). An epidemic model with recruitment–death demographics and discrete delays. In S. Ruan, G. S. K. Wolkowicz, & J. Wu (Eds.), *Differential Equations with Applications to Biology,* pp. 237–49. Providence, RI: American Mathematical Society.

F. R. Giordano & M. D. Weir (1991). *Differential Equations: A Modeling Approach.* Reading, MA: Addison-Wesley.

I. Gladwell (1979a). The development of the boundary-value codes in the ordinary differential equations chapter of the NAG library. In B. Childs et al. (Eds.), *Codes for Boundary Value Problems* (Lecture Notes in Computer Science, 76), pp. 122–43. New York: Springer-Verlag.

I. Gladwell (1979b). Initial value routines in the NAG library. *ACM Trans. Math. Software* 5: 386–400.

I. Gladwell (1987). The NAG library boundary value codes. Numerical Analysis Report no. 134, Department of Mathematics, University of Manchester, U.K.

*H2KL.* The *Harwell 2000 Library,* at ⟨hsl.rl.ac.uk⟩.

E. Hairer, S. P. Nörsett, & G. Wanner (1987). *Solving Ordinary Differential Equations I.* Berlin: Springer-Verlag.

E. Hairer & G. Wanner (1991). *Solving Ordinary Differential Equations II, Stiff and Differential-Algebraic Problems.* Berlin: Springer-Verlag.

J. Hale (1971). *Functional Differential Equations.* Berlin: Springer-Verlag.

P. Henrici (1962). *Discrete Variable Methods in Ordinary Differential Equations.* New York: Wiley.

P. Henrici (1977). *Error Propagation for Difference Methods.* New York: Krieger.

D. J. Higham & N. J. Higham, *MATLAB Guide.* Philadelphia: SIAM.

A. C. Hindmarsh & G. D. Byrne (1976). Applications of EPISODE: An experimental package for the integration of systems of ordinary differential equations. In L. Lapidus & W. E. Schiesser (Eds.), *Numerical Methods for Differential Systems,* pp. 147–66. New York: Academic Press.

M. H. Holmes. *Introduction to Perturbation Methods.* New York: Springer-Verlag.

T. E. Hull, W. H. Enright, B. M. Fellen, & A. E. Sedgwick (1972). Comparing numerical methods for ordinary differential equations. *SIAM J. Numer. Anal.* 9: 603–37.

T. E. Hull, W. H. Enright, & K. R. Jackson (1975). User's guide for DVERK – A subroutine for solving non-stiff ODEs. Report no. 100, Computer Science Department, University of Toronto, Ontario.

IMSL (2002). *The IMSL FORTRAN 77 Mathematics and Statistics Libraries (FNL),* ver. 3.0. Visual Numerics Inc., Houston, TX.

E. Isaacson & H. B. Keller (1966). *Analysis of Numerical Methods.* New York: Wiley.

E. Kamke (1971). *Differentialgleichungen Lösungsmethoden und Lösungen,* vol. I. New York: Chelsea.

H. B. Keller (1992). *Numerical Methods for Two-Point Boundary-Value Problems.* New York: Dover.

J. Kierzenka (1998). Studies in the numerical solution of ordinary differential equations. Doctoral dissertation, Department of Mathematics, Southern Methodist University, Dallas, TX.

J. Kierzenka & L. F. Shampine (2001). A BVP solver based on residual control and the MATLAB PSE. *ACM Trans. Math. Software* 27: 299–316.

H. Koçak (1989). *Differential and Difference Equations through Computer Experiments.* New York: Springer-Verlag.

M. Kubíček, V. Hlaváček, & M. Holodnick (1979). Test examples for comparison of codes for nonlinear boundary value problems in ordinary differential equations. In B. Childs et al. (Eds.), *Codes for Boundary-Value Problems in Ordinary Differential Equations* (Lecture Notes in Computer Science, 76), pp. 325–46. New York: Springer-Verlag.

J. D. Lambert (1991). *Numerical Methods for Ordinary Differential Systems.* New York: Wiley.

L. Lapidus, R. C. Aiken, & Y. A. Liu (1973). The occurrence and numerical solution of physical and chemical systems having widely varying time constants. In R. A. Willoughby (Ed.), *Stiff Differential Systems,* pp. 187–200. New York: Plenum.

H. T. Laquer & B. Wendroff (1981). Bounds for the model quench front. *SIAM J. Numer. Anal.* 18: 225–41.

M. Lentini & V. Pereyra (1974). A variable order finite difference method for nonlinear multipoint boundary value problems. *Math. Comp.* 23: 981–1003.

J. Lighthill (1986). *An Informal Introduction to Theoretical Fluid Mechanics.* Oxford: Clarendon.

C. C. Lin & L. A. Segel (1998). *Mathematics Applied to Deterministic Problems in the Natural Sciences.* Philadelphia: SIAM.

N. MacDonald (1978). *Time Lags in Biological Models.* Berlin: Springer-Verlag.

N. MacDonald (1989). *Biological Delay Systems: Linear Stability Theory.* Cambridge University Press.

Maple (1998). *Maple V Release 6.* Waterloo Maple Inc., Waterloo, Ontario.

M. Marletta & J. D. Pryce (1995). LCNO Sturm–Liouville problems – Computational difficulties and examples. *Numer. Math.* 69: 303–20.

C. Marriott & C. DeLisle (1989). Effects of discontinuities in the behavior of a delay differential equation. *Physica D* 36: 198–206.

A. Martin & S. Ruan (2001). Predator–prey models with delay and prey harvesting. *J. Math. Biol.* 43: 247–67.

MATLAB (2000). *MATLAB 6.* The MathWorks, Inc., Natick, MA.

R. M. M. Mattheij & G. W. M. Staarink (1984a). An efficient algorithm for solving general linear two-point BVP. *SIAM J. Sci. Stat. Comput.* 5: 745–63.

R. M. M. Mattheij & G. W. M. Staarink (1984b). On optimal shooting intervals. *Math. Comp.* 42: 25–40.

C. B. Moler (1997). Are we there yet? *MATLAB Newsletter* (Simulink 2 Special Edition), pp. 16–17; see ⟨http://www.mathworks.com/company/newsletter/pdf/97slCleve.pdf⟩.

C. B. Moler & L. P. Solomon (1970). Integrating square roots. *Comm. ACM* 13: 556–7.

J. S. Murphy (1965). Extensions of the Falkner–Skan similar solutions to flows with surface curvature. *AIAA J.* 3: 2043–9.

J. D. Murray (1993). *Mathematical Biology,* 2nd ed. Berlin: Springer-Verlag.

NAG (2002). *NAG FORTRAN 77 Library,* mark 21. Numerical Algorithms Group Inc., Oxford, U.K.

*Netlib.* The *Netlib* software repository is available at ⟨http://www.netlib.org/⟩.

K. W. Neves (1975). Automatic integration of functional differential equations: An approach. *ACM Trans. Math. Software* 1: 357–68.

K. W. Neves & S. Thompson (1992). Software for the numerical solution of systems of functional differential equations with state dependent delays. *Appl. Numer. Math.* 9: 385–401.

H. J. Oberle & H. J. Pesch (1981). Numerical treatment of delay differential equations by Hermite interpolation. *Numer. Math.* 37: 235–55.

R. E. O'Malley (1991). *Singular Perturbation Methods for Ordinary Differential Equations.* New York: Springer-Verlag.

J. M. Ortega & W. G. Poole (1981). *An Introduction to Numerical Methods for Differential Equations.* Marshfield, MA: Pitman.

J. T. Ottesen (1997). Modelling of the baroflex-feedback mechanism with time-delay. *J. Math. Biol.* 36: 41–63.

C. A. H. Paul (1995). A user-guide to ARCHI. Numerical Analysis Report no. 283, Mathematics Department, University of Manchester, U.K.

S. Pruess, C. T. Fulton, & Y. Xie (1992). Performance of the Sturm–Liouville software SLEDGE. Technical Report no. MCS-91-19, Department of Mathematical Sciences, Colorado School of Mines, Golden.

J. D. Pryce (1993). *Numerical Solution of Sturm–Liouville Problems.* Oxford: Clarendon.

J. D. Pryce (1999). A test package for Sturm–Liouville solvers. *ACM Trans. Math. Software* 25: 21–57.

A. Raghothama & S. Narayanan (2002). Periodic response and chaos in nonlinear systems with parametric excitation and time delay. *Nonlinear Dynam.* 27: 341–65.

S. M. Roberts & J. S. Shipman (1972). *Two-Point Boundary Value Problems: Shooting Methods.* New York: Elsevier.

H. H. Robertson (1996). The solution of a set of reaction rate equations. In J. Walsh (Ed.), *Numerical Analysis: An Introduction*, pp. 178–82. London: Academic Press.

J. M. Sanz-Serna & M. P. Calvo (1994). *Numerical Hamiltonian Problems.* London: Chapman & Hall.

M. R. Scott (1973). *Invariant Imbedding and Its Applications to Ordinary Differential Equations, An Introduction.* Reading, MA: Addison-Wesley.

R. Seydel (1988). *From Equilibrium to Chaos.* New York: Elsevier.

L. F. Shampine (1986). Conservation laws and the numerical solution of ODEs. *Comput Math. Appl.* 12B: 1287–96.

L. F. Shampine (1994). *Numerical Solution of Ordinary Differential Equations.* New York: Chapman & Hall.

L. F. Shampine (1998). Linear conservation laws for ODEs. *Comput Math. Appl.* 35: 45–53.

L. F. Shampine (2002). Variable order Adams codes. *Comput Math. Appl.* 44: 749–61.

L. F. Shampine, R. C. Allen, Jr., & S. Pruess (1997). *Fundamentals of Numerical Computing.* New York: Wiley.

L. F. Shampine, I. Gladwell, & R. W. Brankin (1991). Reliable solution of special root finding problems for ODEs. *ACM Trans. Math. Software* 17: 11–25.

L. F. Shampine & M. K. Gordon (1975). *Computer Solution of Ordinary Differential Equations.* San Francisco: Freeman.

L. F. Shampine & M. W. Reichelt (1997). The MATLAB ODE suite. *SIAM J. Sci. Comput.* 18: 1–22.

L. F. Shampine, M. W. Reichelt, & J. A. Kierzenka (1999). Solving index-1 DAEs in MATLAB and Simulink. *SIAM Review* 41: 538–52.

L. F. Shampine & S. Thompson (2000). Event location for ordinary differential equations. *Comput. Math. Appl.* 39: 43–54.

L. F. Shampine & S. Thompson (2001). Solving DDEs in MATLAB. *Appl. Numer. Math.* 37: 441–58.

R. D. Skeel & M. Berzins (1990). A method for the spatial discretization of parabolic equations in one space variable. *SIAM J. Sci. Stat. Comput.* 11: 1–32.

S. S. Soliman & M. D. Srinath (1998). *Continuous and Discrete Signals and Systems,* 2nd ed. Englewood Cliffs, NJ: Prentice-Hall.

H. J. Stetter (1973). *Analysis of Discretization Methods for Ordinary Differential Equations.* New York: Springer-Verlag.

A. M. Stuart & A. R. Humphries (1996). *Dynamical Systems and Numerical Analysis.* Cambridge University Press.

S. Suherman, R. H. Plaut, L. T. Watson, & S. Thompson (1997). Effect of human response time on rocking instability of a two-wheeled suitcase. *J. Sound Vibration* 207: 617–25.

L. Tavernini (1996). *Continuous-Time Modeling and Simulation*. Amsterdam: Gordon & Breach.

S. Thompson & P. G. Tuttle (1986). Benchmark fluid flow problems for continuous simulation languages. *Comput. Math. Appl.* 12A: 345–52.

L. N. Trefethen (2000). *Spectral Methods in MATLAB*. Philadelphia: SIAM.

D. A. Wells (1967). *Theory and Problems of Lagrangian Dynamics* (Schaum's Outline Series). New York: McGraw-Hill.

D. R. Willé & C. T. H. Baker (1992). DELSOL – A numerical code for the solution of systems of delay-differential equations. *Appl. Numer. Math.* 9: 223–34.

S. J. Wolfram (1996). *The Mathematica Book*, 3rd ed. Wolfram Media & Cambridge University Press.

# Index

CL

515.
35
SHA

6000770523

BIB 441852.